人工智能与大数据系列

OpenCV入门与技术实践

罗刚　编著

清华大学出版社
北京

内容简介

本书介绍如何学习和使用流行的 OpenCV 库开发计算机视觉应用。主要内容包括图像的核心操作、图像阈值处理、图像形态变换、图像边缘检测、角点检测与特征匹配等。

全书分为 4 章：第 1 章着重介绍使用 Python 开发 OpenCV 应用基础知识；第 2 章着重介绍使用 OpenCV 进行图像特征检测、描述和特征匹配的各种算法；第 3 章着重介绍 OCR 文字识别；第 4 章着重介绍 OpenCV 中的深度学习。

本书适合作为高等院校计算机、软件工程专业本科生、研究生的参考书目，也适用于对人工智能领域感兴趣的人士。

本书封面贴有清华大学出版社防伪标签，无标签者不得销售。

版权所有，侵权必究。举报：010-62782989，beiqinquan@tup.tsinghua.edu.cn。

图书在版编目（CIP）数据

OpenCV 入门与技术实践 / 罗刚编著. —北京：清华大学出版社，2023.7
（人工智能与大数据系列）
ISBN 978-7-302-63224-5

Ⅰ. ①O… Ⅱ. ①罗… Ⅲ. ①图像处理软件－程序设计 Ⅳ. ①TP391.413

中国国家版本馆 CIP 数据核字（2023）第 056665 号

责任编辑：张　敏
封面设计：郭二鹏
责任校对：胡伟民
责任印制：丛怀宇

出版发行：清华大学出版社
网　　址：http://www.tup.com.cn, http://www.wqbook.com
地　　址：北京清华大学学研大厦 A 座　　邮　编：100084
社　总　机：010-83470000　　邮　购：010-62786544
投稿与读者服务：010-62776969, c-service@tup.tsinghua.edu.cn
质量反馈：010-62772015, zhiliang@tup.tsinghua.edu.cn
课件下载：http://www.tup.com.cn, 010-83470236

印装者：三河市科茂嘉荣印务有限公司
经　　销：全国新华书店
开　　本：185mm×260mm　　印　张：9.25　　字　数：252 千字
版　　次：2023 年 9 月第 1 版　　印　次：2023 年 9 月第 1 次印刷
定　　价：58.00 元

产品编号：087125-01

前言

OpenCV 是一个主要针对实时计算机视觉的编程函数库。OpenCV 支持 Python、C++、Java 等多种编程语言。本书介绍使用 Python 开发 OpenCV 应用。

全书共 4 章：第 1 章介绍使用 Python 开发 OpenCV 应用基础知识；第 2 章介绍使用 OpenCV 进行图像特征检测、描述和特征匹配的各种算法；第 3 章介绍 OCR 文字识别；第 4 章介绍 OpenCV 中的深度学习。

本书适合需要具体实现计算机视觉应用的开发人员或者对人工智能等相关领域感兴趣的人士参考。同时，猎兔搜索技术团队已经开发出与本书配套的培训课程和商业软件。

感谢早期合著者、合作伙伴、员工、学员、读者的支持，给我们提供了良好的工作基础。就像玻璃容器中的水培植物一样，这是一个持久可用的成长基础。技术的融合与创新无止境，欢迎一起探索。

目录
CONTENTS

第 1 章　使用 Python 开发 OpenCV 的应用 ···1
 1.1　准备工作环境 ···1
 1.2　安装 ···2
  1.2.1　检测 OpenCV 版本 ··2
  1.2.2　读取、显示和写入图像文件示例 ···2
 1.3　视频入门 ···3
 1.4　绘图函数 ···5
 1.5　图像的核心操作 ···6
  1.5.1　基本操作 ··6
  1.5.2　图像上的算术运算 ··7
 1.6　改变颜色空间 ···8
 1.7　图像变换 ···9
 1.8　图像阈值处理 ···12
  1.8.1　简单阈值化 ··12
  1.8.2　自适应阈值化 ··13
  1.8.3　Otsu二值化 ··13
 1.9　平滑图像 ···14
 1.10　形态变换 ···16
 1.11　斑点检测 ···17
 1.12　Sobel 边缘检测方法 ··20
 1.13　Canny 边缘检测 ···21
 1.14　轮廓检测 ···22
 1.15　人脸检测 ···29
 1.16　特征脸 ···30
 1.17　图像金字塔 ···34
 1.18　实时人体检测 ···35
 1.19　背景减除 ···37
 1.20　模板匹配 ···38
 1.21　直线检测 ···39

1.22 霍夫圆变换 40
1.23 镜头畸变 41
1.24 使用 Hu 矩进行形状匹配 42
1.25 找到 blob 的中心 44
1.26 查找凸包 46
1.27 将一个三角形扭曲为另一个三角形 47
1.28 阿尔法混合 49
1.29 基于特征的图像对齐 50
1.30 使用 ZBar 和 OpenCV 编写条形码和二维码扫描仪的 Python 代码 52
1.31 换脸 53
1.32 applyColorMap 用于伪着色 56
1.33 高动态范围成像 57
1.34 曝光融合 60
1.35 对象跟踪 63
1.36 多对象跟踪 66
1.37 自动红眼去除器 69
1.38 创建虚拟笔和橡皮擦 71
1.39 使用 ArUco 标记的增强现实 85

第 2 章 特征检测和描述 89

2.1 Harris 角点检测 89
2.2 Shi-Tomasi 角点检测器和良好的跟踪功能 90
2.3 尺度不变特征变换 90
2.4 特征匹配 91
2.5 特征匹配+单应性查找对象 93

第 3 章 OCR 文字识别 95

3.1 OpenCV 的 OCR 功能 95
3.2 Tesseract 的预处理 100

第 4 章 OpenCV 深度学习 102

4.1 使用 OpenCV DNN 模块进行深度学习 102
4.2 基于 OpenCV 和深度学习的人脸识别 109
4.3 英特尔 OpenVINO 工具包简介 112
4.4 使用深度学习的自动车牌识别 117
    4.4.1 使用 YOLOv4 检测车牌 118
    4.4.2 OCR 121
4.5 超分辨率 123
4.6 对象检测 127
4.7 GOTURN：基于深度学习的对象跟踪 131
4.8 手势识别 134
4.9 人体姿态估计 136
4.10 使用 OpenPose 在 OpenCV 中进行多人姿态估计 139

第1章
使用 Python 开发 OpenCV 的应用

在人工智能领域，计算机视觉是最有趣和最具挑战性的任务之一。计算机视觉就像计算机软件和我们周围的可视化之间的桥梁。它允许计算机软件理解和了解周围环境的可视化。例如，根据颜色、形状和大小确定水果的种类。这项任务对人脑来说非常容易，但是在计算机视觉管道中，首先要收集数据，其次进行数据处理活动，然后训练和教授模型以使其了解如何根据大小、形状和颜色区分水果。

目前，已有多种包可执行机器学习、深度学习和计算机视觉任务。OpenCV 是一个开源库，支持 R、Python、C++、Java 等多种编程语言。OpenCV 可以在大多数平台上运行，如 Windows、Linux 和 macOS。

1.1 准备工作环境

首先要准备一个 Python 的开发环境。当前可以使用 Python 3.9 版本。Python 3.9 可以从 Python 的官方网站 https://www.python.org/下载得到。使用默认方式安装即可。

在 Windows 下安装 Python 以后，在控制台输入 python 命令进入交互式环境。

```
C:\Users\Administrator>python
Python 3.9.6 (tags/v3.9.6:db3ff76, Jun 28 2021, 15:26:21) [MSC v.1929 64 bit (AMD64)] on win32
Type "help", "copyright", "credits" or "license" for more information
>>>
```

在 Windows 操作系统下，检查 Python 3 是否已经正确安装，以及安装的版本号：

```
>python3 -V
Python 3.9.6
```

检查 Python 3 所在的路径：

```
>where python3
C:\Users\Administrator\AppData\Local\Microsoft\WindowsApps\python3.exe
```

可以准备一个用于编写代码的集成开发环境。例如，可以使用 PyCharm 或者 Visual Studio，也可以简单使用 Notepad++这样的文本编辑器编写 Python 代码。

1.2 安装

使用清华大学的镜像作为 pip 源。

```
>pip config set global.index-url https://pypi.tuna.tsinghua.edu.cn/simple
```

安装 opencv-contrib-python 模块：

```
>pip install opencv-contrib-python
```

1.2.1 检测 OpenCV 版本

cv2.__version__ 提供了版本字符串。可以从中提取主要和次要版本，如下所示。

```python
import cv2
# Print version string
print( "OpenCV version : {0}".format(cv2.__version__))

# Extract major, minor, and subminor version numbers
(major_ver, minor_ver, subminor_ver) = (cv2.__version__).split('.')
print( "Major version : {0}".format(major_ver))
print( "Minor version : {0}".format(minor_ver))
print( "Submior version : {0}".format(subminor_ver))
```

1.2.2 读取、显示和写入图像文件示例

读取、显示和写入图像是图像处理和计算机视觉的基础。
读取图像：

```python
# Importing the OpenCV library
import cv2
# Reading the image using imread() function
image = cv2.imread('image.png')

# Extracting the height and width of an image
h, w = image.shape[:2]
# Displaying the height and width
print("Height = {}, Width = {}".format(h, w))
```

现在我们将专注于提取单个像素的 RGB 值。
注意：OpenCV 按 BGR 顺序排列通道。所以第 0 个值将对应于蓝色像素而不是红色像素。
如下代码提取像素的 BGR 值：

```python
# Extracting RGB values.
# Here we have randomly chosen a pixel
# by passing in 100, 100 for height and width
(B, G, R) = image[100, 100]
```

```
# Displaying the pixel values
print("R = {}, G = {}, B = {}".format(R, G, B))

# We can also pass the channel to extract
# the value for a specific channel
B = image[100, 100, 0]
print("B = {}".format(B))
```

提取感兴趣区域（ROI）：

```
# We will calculate the region of interest
# by slicing the pixels of the image
roi = image[100 : 500, 200 : 700]
```

显示图像：

```
#Displays image inside a window
cv2.imshow('color image',img_color)
cv2.imshow('grayscale image',img_grayscale)
cv2.imshow('unchanged image',img_unchanged)

# Waits for a keystroke
cv2.waitKey(0)

# Destroys all the windows created
cv2.destroyAllwindows()
```

显示文字是一个就地操作：

```
# Copying the original image
output = image.copy()

# Adding the text using putText() function
text = cv2.putText(output, 'OpenCV Demo', (500, 550),
            cv2.FONT_HERSHEY_SIMPLEX, 4, (255, 0, 0), 2)
```

cv2.putText()函数接收 7 个参数：
（1）图像。
（2）要显示的文本。
（3）左下角坐标，文本应该从哪里开始出现的位置。
（4）字体。
（5）字体大小。
（6）颜色（RGB 格式）。
（7）行宽。
写图像：

```
cv2.imwrite('grayscale.jpg',img_grayscale)
```

1.3 视频入门

通常，捕捉实时流必须用相机。OpenCV 提供了一个非常简单的接口来执行此操作，让我们从相机中捕捉视频，将其转换为灰度视频并显示。

为了捕获视频，需要创建一个 VideoCapture 对象。它的参数可以是设备索引或视频文件的名称。设备索引只是指定哪个相机的数字，因为通常会连接一台相机（例如笔记本计算机上的内置网络摄像头），所以这里只是传递 0（或-1）。也可以通过传递 1 来选择第二个相机，依此类推。之后，可以逐帧捕获，但最后不要忘记释放捕获。

```python
import numpy as np
import cv2 as cv
cap = cv.VideoCapture(0)
if not cap.isOpened():
    print("Cannot open camera")
    exit()
while True:
    # Capture frame-by-frame
    ret, frame = cap.read()
    # if frame is read correctly ret is True
    if not ret:
        print("Can't receive frame (stream end?). Exiting ...")
        break
    # Our operations on the frame come here
    gray = cv.cvtColor(frame, cv.COLOR_BGR2GRAY)
    # Display the resulting frame
    cv.imshow('frame', gray)
    if cv.waitKey(1) == ord('q'):
        break
# When everything done, release the capture
cap.release()
cv.destroyAllWindows()
```

cap.read()返回一个布尔值（真/假）。如果帧被正确读取，它将是 True。因此，可以通过检查此返回值来检查视频的结束。

有时，cap 可能没有初始化捕获。在这种情况下，此代码显示错误。可以通过 cap.isOpened() 方法检查它是否被初始化。

还可以使用 cap.get(propId)方法访问此视频的某些功能，其中 propId 是从 0 到 18 的数字。每个数字表示视频的一个属性（如果它适用于该视频）。其中一些值可以使用 cap.set(propId, value) 进行修改。这里的 value 是想要的新值。

例如，可以通过 cap.get(cv.CAP_PROP_FRAME_WIDTH)和 cap.get(cv.CAP_PROP_FRAME_HEIGHT)检查帧的宽度和高度。默认情况下返回值是 640×480。但如果想将其修改为 320×240。只需使用 ret=cap.set(cv.CAP_PROP_FRAME_WIDTH, 320)和 ret= cap.set(cv.CAP_PROP_FRAME_HEIGHT, 240)。

从文件播放视频与从相机捕获视频相同，只须将相机索引更改为视频文件名即可。

```python
import numpy as np
import cv2 as cv
cap = cv.VideoCapture('vtest.avi')
while cap.isOpened():
    ret, frame = cap.read()
    # if frame is read correctly ret is True
```

```
        if not ret:
            print("Can't receive frame (stream end?). Exiting ...")
            break
        gray = cv.cvtColor(frame, cv.COLOR_BGR2GRAY)
        cv.imshow('frame', gray)
        if cv.waitKey(1) == ord('q'):
            break
    cap.release()
    cv.destroyAllWindows()
```

接下来捕获一个视频并逐帧处理它，我们想要保存该视频。下面的代码从相机捕获一个视频，在垂直方向进行逐帧翻转，并保存视频。

```
    import numpy as np
    import cv2 as cv
    cap = cv.VideoCapture(0)
    # Define the codec and create VideoWriter object
    fourcc = cv.VideoWriter_fourcc(*'XVID')
    out = cv.VideoWriter('output.avi', fourcc, 20.0, (640, 480))
    while cap.isOpened():
        ret, frame = cap.read()
        if not ret:
            print("Can't receive frame (stream end?). Exiting ...")
            break
        frame = cv.flip(frame, 0)
        # write the flipped frame
        out.write(frame)
        cv.imshow('frame', frame)
        if cv.waitKey(1) == ord('q'):
            break
    # Release everything if job is finished
    cap.release()
    out.release()
    cv.destroyAllWindows()
```

1.4 绘图函数

要绘制一条线，需要传递线的起点坐标和终点坐标。我们将创建一个黑色图像并在其上从左上角到右下角画一条蓝线。

```
    import numpy as np
    import cv2 as cv
    # Create a black image
    img = np.zeros((512,512,3), np.uint8)
    # Draw a diagonal blue line with thickness of 5 px
    cv.line(img,(0,0),(511,511),(255,0,0),5)
```

要绘制矩形，需要给出矩形的左上角和右下角坐标。这次我们将在图像的右上角绘制一个绿色矩形。

```
cv.rectangle(img,(384,0),(510,128),(0,255,0),3)
```

要画一个圆,需要它的中心坐标和半径。我们将在上面绘制的矩形内画一个圆圈。

```
cv.circle(img,(447,63), 63, (0,0,255), -1)
```

要绘制椭圆,需要传递几个参数。一个参数是中心位置 (x, y)。下一个参数是轴长度(长轴长度,短轴长度)。angle 是椭圆沿逆时针方向旋转的角度。startAngle 和 endAngle 表示从主轴顺时针方向测量的椭圆弧的起点和终点,即给出值 0 和 360 画出完整的椭圆。下面的示例为在图像的中心绘制一个半椭圆。

```
cv.ellipse(img,(256,256),(100,50),0,0,180,255,-1)
```

要绘制多边形,首先需要知道顶点坐标。将这些点放入一个形状为 ROWS×1×2 的数组中,其中 ROWS 是顶点数,它应该是 int32 类型。在这里,我们用黄色绘制一个带有 4 个顶点的小多边形。

```
pts = np.array([[10,5],[20,30],[70,20],[50,10]], np.int32)
pts = pts.reshape((-1,1,2))
cv.polylines(img,[pts],True,(0,255,255))
```

1.5 图像的核心操作

首先介绍图像上的基本操作,然后介绍图像上的算术运算。

1.5.1 基本操作

为了访问和修改像素值,先加载一幅彩色图像:

```
>>> import numpy as np
>>> import cv2 as cv
>>> img = cv.imread('messi5.jpg')
```

可以通过其行和列坐标访问像素值。对于 BGR 图像,它返回一个包含蓝色、绿色、红色值的数组。对于灰度图像,只返回相应的强度。

```
>>> px = img[100,100]
>>> print( px )
[157 166 200]
# accessing only blue pixel
>>> blue = img[100,100,0]
>>> print( blue )
157
```

可以以相同的方式修改像素值。

```
>>> img[100,100] = [255,255,255]
>>> print( img[100,100] )
[255 255 255]
```

更好的像素访问和编辑方法:

```
# accessing RED value
>>> img.item(10,10,2)
```

```
59
# modifying RED value
>>> img.itemset((10,10,2),100)
>>> img.item(10,10,2)
100
```

图像属性包括行数、列数和通道数，图像数据的类型，像素数，等等。

通过 img.shape 访问图像的形状。它返回一个包含行数、列数和通道数的元组（如果图像是彩色的）：

```
>>> print( img.shape )
(342, 548, 3)
```

通过 img.size 访问像素总数：

```
>>> print( img.size )
562248
```

图像数据类型通过 img.dtype 获得：

```
>>> print( img.dtype )
uint8
```

有时，将不得不使用某些图像区域。对于图像中的眼睛检测，首先对整幅图像进行人脸检测。当获得人脸时，我们只选择人脸区域并在其中搜索眼睛，而不是搜索整幅图像。这样提高了准确性（因为眼睛总是在脸上）和性能（因为我们在一个小区域内搜索）。

使用 NumPy 索引获得 ROI。在这里，选择了一个球并将其复制到图像中的另一个区域：

```
>>> ball = img[280:340, 330:390]
>>> img[273:333, 100:160] = ball
```

有时需要单独处理图像的 B、G、R 通道。在这种情况下，需要将 BGR 图像拆分为单个通道。在其他情况下，可能需要加入这些单独的频道来创建 BGR 图像。可以通过以下方式简单地做到这一点：

```
>>> b,g,r = cv.split(img)
>>> img = cv.merge((b,g,r))
```

假设想将所有红色像素设置为 0，不需要先拆分通道，而使用 NumPy 索引更快：

```
>>> img[:,:,2] = 0
```

1.5.2 图像上的算术运算

可以使用 OpenCV 函数 cv.add()或简单地通过 NumPy 操作 res = img1 + img2 相加两幅图像。两幅图像应该具有相同的深度和类型，或者第二幅图像可以只是一个标量值。

例如，考虑以下示例：

```
>>> x = np.uint8([250])
>>> y = np.uint8([10])
>>> print( cv.add(x,y) )   # 250+10 = 260 => 255
[[255]]
>>> print( x+y )           # 250+10 = 260 % 256 = 4
[4]
```

为图像赋予不同的权重，以产生混合或透明的感觉也是图像相加。根据以下等式相加图像：

$$g(x)=(1-\alpha)f_0(x)+\alpha f_1(x)$$

通过将 α 从 0→1 变化，可以在一幅图像到另一幅图像之间执行一个很酷的过渡。

在这里，将两幅图像混合在一起。第一幅图像的权重为 0.7，第二幅图像的权重为 0.3。通过 cv.addWeighted() 将以下等式应用于图像：

$$dst=\alpha \cdot img1+\beta \cdot img2+\gamma$$

这里 γ 取值为 0。

```
img1 = cv.imread('ml.png')
img2 = cv.imread('opencv-logo.png')
dst = cv.addWeighted(img1,0.7,img2,0.3,0)
cv.imshow('dst',dst)
cv.waitKey(0)
cv.destroyAllWindows()
```

位运算包括按位与、或、非和异或运算。它们在提取图像的任何部分、定义和使用非矩形 ROI 时非常有用。下面我们将看到一个如何更改图像特定区域的示例。

想将 OpenCV 徽标放在图像上方。如果添加两幅图像，它会改变颜色。如果混合它们，会得到透明的效果。但希望它是不透明的。如果它是一个矩形区域，可以像上一节那样使用 ROI。但是 OpenCV 标志不是一个矩形。因此，可以使用按位运算来完成，如下所示：

```
# Load two images
img1 = cv.imread('messi5.jpg')
img2 = cv.imread('opencv-logo-white.png')
# I want to put logo on top-left corner, So I create a ROI
rows,cols,channels = img2.shape
roi = img1[0:rows, 0:cols]
# Now create a mask of logo and create its inverse mask also
img2gray = cv.cvtColor(img2,cv.COLOR_BGR2GRAY)
ret, mask = cv.threshold(img2gray, 10, 255, cv.THRESH_BINARY)
mask_inv = cv.bitwise_not(mask)
# Now black-out the area of logo in ROI
img1_bg = cv.bitwise_and(roi,roi,mask = mask_inv)
# Take only region of logo from logo image.
img2_fg = cv.bitwise_and(img2,img2,mask = mask)
# Put logo in ROI and modify the main image
dst = cv.add(img1_bg,img2_fg)
img1[0:rows, 0:cols ] = dst
cv.imshow('res',img1)
cv.waitKey(0)
cv.destroyAllWindows()
```

1.6 改变颜色空间

OpenCV 中使用最广泛的两种颜色空间转换方法是 BGR↔GRAY（灰度）和 BGR↔HSV。颜色空间转换可以使用函数 cv.cvtColor(input_image, flag)，其中 flag 决定转换的类型。

对于 BGR→灰度颜色空间的转换，可以使用标志 cv.COLOR_BGR2GRAY。类似地，对于

BGR→HSV 颜色空间的转换，可以使用标志 cv.COLOR_BGR2HSV。要获取其他标志，只须在 Python 终端运行以下命令：

```
>>> import cv2 as cv
>>> flags = [i for i in dir(cv) if i.startswith('COLOR_')]
>>> print( flags )
```

前面已经介绍了如何将 BGR 颜色空间的图像转换为 HSV 颜色空间的图像，现在可以使用它来提取彩色对象。在 HSV 颜色空间表示颜色比在 BGR 颜色空间更容易。在下面的应用程序中，将尝试提取蓝色对象。方法如下：

（1）逐帧取视频。
（2）从 BGR 颜色空间转换到 HSV 颜色空间。
（3）对 HSV 颜色空间的图像设置蓝色范围的阈值。
（4）单独提取蓝色对象，可以对图像做任何想做的事情。

下面是详细注释的代码：

```
import cv2 as cv
import numpy as np
cap = cv.VideoCapture(0)
while(1):
    # Take each frame
    _, frame = cap.read()
    # Convert BGR to HSV
    hsv = cv.cvtColor(frame, cv.COLOR_BGR2HSV)
    # define range of blue color in HSV
    lower_blue = np.array([110,50,50])
    upper_blue = np.array([130,255,255])
    # Threshold the HSV image to get only blue colors
    mask = cv.inRange(hsv, lower_blue, upper_blue)
    # Bitwise-AND mask and original image
    res = cv.bitwise_and(frame,frame, mask= mask)
    cv.imshow('frame',frame)
    cv.imshow('mask',mask)
    cv.imshow('res',res)
    k = cv.waitKey(5) & 0xFF
    if k == 27:
        break
cv.destroyAllWindows()
```

1.7 图像变换

本节讲解使用平移、旋转、调整大小、翻转和裁剪等来修改图像的各种方法。
首先，导入用于转换图像的模块。

```
# importing the numpy module to work with pixels in images
import numpy as np

# importing argument parsers
```

```python
import argparse

# importing the OpenCV module
import cv2
```

接下来，设置参数解析器，以便可以从用户那里获取有关图像文件位置的输入。

```python
# initializing an argument parser object
ap = argparse.ArgumentParser()

# adding the argument, providing the user an option
# to input the path of the image
ap.add_argument("-i", "--image", required=True, help="Path to the image")

# parsing the argument
args = vars(ap.parse_args())
```

在 OpenCV 中使用平移，首先要定义一个接收输入图像以及沿 x 轴和 y 轴平移的函数。warpAffine()函数既接收输入图像，也接收平移矩阵，并在平移过程中弯曲图像。

最后，将更改后的图像返回给程序。

```python
# defining a function for translation
def translate(image, x, y):
    # defining the translation matrix
    M = np.float32([[1, 0, x], [0, 1, y]])

    # the cv2.warpAffine method does the actual translation
    # containing the input image and the translation matrix
    shifted = cv2.warpAffine(image, M, (image.shape[1], image.shape[0]))

    # we then return the image
    return shifted
```

现在，已有了负责接收输入和提供输出的代码。translate()函数提供了对流程的简单直接调用。

```python
# reads the image from image location
image = cv2.imread(args["image"])
cv2.imshow("Original", image)

# call the translation function to translate the image
shifted = translate(image, 0, 100)
cv2.imshow("Shifted Down", shifted)
cv2.waitKey(0)
```

此时应该会收到一幅与原始图像一起向下移动 100 像素的图像。

下面首先定义一个旋转函数，以便稍后使用一行代码来旋转图像。

rotate()函数接收图像、必须旋转图像的角度，因此还将声明中心和缩放的默认值。

cv2.getRotationMatrix2D()可以创建一个矩阵，该矩阵在变形时提供旋转图像。

然后，返回旋转后的图像。

```python
# defining a function for rotation
def rotate(image, angle, center=None, scale=1.0):
    (h, w) = image.shape[:2]
```

```
    if center is None:
        center = (w / 2, h / 2)

    # the cv2.getRotationMatrix2D allows us to create a
    # Rotation matrix
    M = cv2.getRotationMatrix2D(center, angle, scale)

    # the warpAffine function allows us to rotate the image
    # using the rotation matrix
    rotated = cv2.warpAffine(image, M, (w, h))

    return rotated
```

现在，让我们通过为图像提供不同的角度来测试该函数。

```
# rotating the image by 45 degrees
rotated = rotate(image, 45)
cv2.imshow("Rotated by 45 Degrees", rotated)

# rotating the image by 90 degrees
rotated = rotate(image, 90)
cv2.imshow("Rotated by -90 Degrees", rotated)

# rotating the image by 180 degrees
rotated = rotate(image, 180)
cv2.imshow("Rotated by 180 degrees", rotated)
```

OpenCV 中的翻转非常简单，只需要一个简单的 flip() 函数即可。

cv2.flip() 函数接收两个参数：一个是图像本身；另一个表示如何翻转图像。翻转参数表如表 1-1 所示。

表 1-1 翻转参数表

参 数 值	如何翻转图像
0	垂直翻转
1	水平翻转
-1	垂直和水平翻转

翻转图像的代码如下：

```
# flipping the image horizontally
flipped = cv2.flip(image, 1)
cv2.imshow("Flipped Horizontally", flipped)

# flipping the image vertically
flipped = cv2.flip(image, 0)
cv2.imshow("Flipped Vertically", flipped)

# flipping the image vertically and horizontally
flipped = cv2.flip(image, -1)
cv2.imshow("Flipped Vertically and Horizontally", flipped)
```

```python
# wait for the user's key to proceed
cv2.waitKey(0)
```

在 cv2 中裁剪图像就像在 Python 中访问列表一样简单。没有关于它的函数,因为不需要这样一个函数。

裁剪图像的代码如下:

```python
# displaying the width and height of the image
print("Width", image.shape[1])
print("Height", image.shape[0])

# cropping the image manually
face = image[:400, :600]
cv2.imshow("Face", face)
cv2.waitKey(0)

# cropping the image manually
body = image[400:600, 120:600]
cv2.imshow("Body", body)
cv2.waitKey(0)
```

1.8 图像阈值处理

本节讲解简单阈值化、自适应阈值化和 Otsu 阈值化。

1.8.1 简单阈值化

简单阈值化是直截了当的,即对于所有像素都应用相同的阈值。如果像素值小于阈值,则设置为 0,否则设置为最大值。函数 cv.threshold()用于应用阈值。函数的第一个参数是源图像,应该是灰度图像;第二个参数是用于对像素值进行分类的阈值;第三个参数是分配给超过阈值的像素值的最大值。OpenCV 提供了不同类型的阈值化,由函数的第四个参数给出。如上所述的基本阈值化处理是通过使用类型 cv.THRESH_BINARY 完成的。所有简单的阈值化类型有:

- cv.THRESH_BINARY。
- cv.THRESH_BINARY_INV。
- cv.THRESH_TRUNC。
- cv.THRESH_TOZERO。
- cv.THRESH_TOZERO_INV。

该方法返回两个输出:第一个是使用的阈值;第二个输出是阈值图像。

如下代码比较了不同的简单阈值化类型:

```python
import cv2 as cv
import numpy as np
from matplotlib import pyplot as plt
img = cv.imread('gradient.png',0)
ret,thresh1 = cv.threshold(img,127,255,cv.THRESH_BINARY)
ret,thresh2 = cv.threshold(img,127,255,cv.THRESH_BINARY_INV)
```

```
ret,thresh3 = cv.threshold(img,127,255,cv.THRESH_TRUNC)
ret,thresh4 = cv.threshold(img,127,255,cv.THRESH_TOZERO)
ret,thresh5 = cv.threshold(img,127,255,cv.THRESH_TOZERO_INV)
titles = ['Original Image','BINARY','BINARY_INV','TRUNC','TOZERO','TOZERO_INV']
images = [img, thresh1, thresh2, thresh3, thresh4, thresh5]
for i in range(6):
    plt.subplot(2,3,i+1),plt.imshow(images[i],'gray',vmin=0,vmax=255)
    plt.title(titles[i])
    plt.xticks([]),plt.yticks([])
plt.show()
```

1.8.2 自适应阈值化

1.8.1 节，使用了一个全局值作为阈值，但这可能并不适用于所有情况，例如，如果图像在不同区域具有不同的照明条件。在这种情况下，自适应阈值化可以提供帮助。在这里，自适应阈值算法根据像素周围的小区域确定像素的阈值，因此可为同一图像在不同区域获得不同的阈值，这对于具有不同照明条件的图像提供了更好的结果。

除了上述参数之外，方法 cv.adaptiveThreshold 还接收以下 3 个输入参数。

（1）参数 AdaptiveMethod 决定如何计算阈值：
- cv.ADAPTIVE_THRESH_MEAN_C：阈值是邻域面积的平均值减去常数 C。
- cv.ADAPTIVE_THRESH_GAUSSIAN_C：阈值是邻域值的高斯加权和减去常数 C。

（2）参数 blockSize 确定邻域区域的大小。

（3）参数 C 是从邻域像素的平均值或加权和中减去的常数。

下面的代码比较了具有不同照明的图像的全局阈值化和自适应阈值化：

```
import cv2 as cv
import numpy as np
from matplotlib import pyplot as plt
img = cv.imread('sudoku.png',0)
img = cv.medianBlur(img,5)
ret,th1 = cv.threshold(img,127,255,cv.THRESH_BINARY)
th2 = cv.adaptiveThreshold(img,255,cv.ADAPTIVE_THRESH_MEAN_C,\
        cv.THRESH_BINARY,11,2)
th3 = cv.adaptiveThreshold(img,255,cv.ADAPTIVE_THRESH_GAUSSIAN_C,\
        cv.THRESH_BINARY,11,2)
titles = ['Original Image', 'Global Thresholding (v = 127)',
        'Adaptive Mean Thresholding', 'Adaptive Gaussian Thresholding']
images = [img, th1, th2, th3]
for i in range(4):
    plt.subplot(2,2,i+1),plt.imshow(images[i],'gray')
    plt.title(titles[i])
    plt.xticks([]),plt.yticks([])
plt.show()
```

1.8.3 Otsu 二值化

在全局阈值中，可以使用任意选择的值作为阈值。相比之下，Otsu（最大类间方差）方法避免了必须选择一个值并自动确定它。

考虑只有两个不同图像值的图像（双峰图像），其中直方图仅包含两个峰值。一个好的阈值将在这两个值的中间。类似地，Otsu 方法从图像直方图中确定了一个最佳的全局阈值。

为此，使用了 cv.threshold() 函数，其中 cv.THRESH_OTSU 作为额外标志传递。阈值可以任意选择。该算法然后找到作为第一个输出返回的最佳阈值。

通过下面的例子，可以了解噪声过滤如何改善结果。输入图像是噪声图像，在第一种情况下，应用值为 127 的全局阈值；在第二种情况下，直接应用 Otsu 阈值；在第三种情况下，首先使用 5×5 高斯核对图像进行过滤以去除噪声，然后应用 Otsu 阈值。

```python
import cv2 as cv
import numpy as np
from matplotlib import pyplot as plt
img = cv.imread('noisy2.png',0)
# global thresholding
ret1,th1 = cv.threshold(img,127,255,cv.THRESH_BINARY)
# Otsu's thresholding
ret2,th2 = cv.threshold(img,0,255,cv.THRESH_BINARY+cv.THRESH_OTSU)
# Otsu's thresholding after Gaussian filtering
blur = cv.GaussianBlur(img,(5,5),0)
ret3,th3 = cv.threshold(blur,0,255,cv.THRESH_BINARY+cv.THRESH_OTSU)
# plot all the images and their histograms
images = [img, 0, th1,
          img, 0, th2,
          blur, 0, th3]
titles = ['Original Noisy Image','Histogram','Global Thresholding (v=127)',
          'Original Noisy Image','Histogram',"Otsu's Thresholding",
          'Gaussian filtered Image','Histogram',"Otsu's Thresholding"]
for i in range(3):
    plt.subplot(3,3,i*3+1),plt.imshow(images[i*3],'gray')
    plt.title(titles[i*3]), plt.xticks([]), plt.yticks([])
    plt.subplot(3,3,i*3+2),plt.hist(images[i*3].ravel(),256)
    plt.title(titles[i*3+1]), plt.xticks([]), plt.yticks([])
    plt.subplot(3,3,i*3+3),plt.imshow(images[i*3+2],'gray')
    plt.title(titles[i*3+2]), plt.xticks([]), plt.yticks([])
plt.show()
```

1.9 平滑图像

与一维信号一样，图像也可以使用各种低通滤波器（LPF）、高通滤波器（HPF）等进行滤波。其中，LPF 有助于去除噪声、模糊图像等；HPF 有助于在图像中寻找边缘。

OpenCV 提供了一个函数 cv.filter2D() 来将内核与图像进行卷积。例如，在图像上尝试平均滤波器，可试试下面的代码并检查结果：

```python
import numpy as np
import cv2 as cv
from matplotlib import pyplot as plt
```

```
img = cv.imread('opencv_logo.png')
kernel = np.ones((5,5),np.float32)/25
dst = cv.filter2D(img,-1,kernel)
plt.subplot(121),plt.imshow(img),plt.title('Original')
plt.xticks([]), plt.yticks([])
plt.subplot(122),plt.imshow(dst),plt.title('Averaging')
plt.xticks([]), plt.yticks([])
plt.show()
```

图像模糊是通过将图像与低通滤波器内核进行卷积来实现的。它对于消除噪声很有用。它实际上从图像中去除了高频内容（如噪声、边缘）。所以，在这个操作中边缘会有点模糊（也有不模糊边缘的模糊技术）。OpenCV 提供了 4 种主要类型的模糊技术。

1. 平均

这是通过将图像与归一化框滤波器进行卷积来完成的。它只是取内核区域下所有像素的平均值并替换中心元素。这是通过函数 cv.blur()或 cv.boxFilter()完成的。下面的示例演示使用 5×5 大小的内核：

```
import cv2 as cv
import numpy as np
from matplotlib import pyplot as plt
img = cv.imread('opencv-logo-white.png')
blur = cv.blur(img,(5,5))
plt.subplot(121),plt.imshow(img),plt.title('Original')
plt.xticks([]), plt.yticks([])
plt.subplot(122),plt.imshow(blur),plt.title('Blurred')
plt.xticks([]), plt.yticks([])
plt.show()
```

2. 高斯模糊

在此方法中，使用高斯核代替盒式滤波器。这是通过函数 cv.GaussianBlur()完成的。在其过程中，应该指定内核的宽度和高度，它们应该是正数和奇数；还应该分别指定 X 和 Y 方向的标准偏差 sigmaX 和 sigmaY。如果仅指定 sigmaX，则 sigmaY 与 sigmaX 相同；如果两者都为零，则它们是根据内核大小计算的。高斯模糊对于从图像中去除高斯噪声非常有效。

如果需要，可以使用函数 cv.getGaussianKernel()创建高斯核。

上面的代码可以修改为高斯模糊：

```
blur = cv.GaussianBlur(img,(5,5),0)
```

3. 中值模糊

这里，函数 cv.medianBlur()取内核区域下所有像素的中值，并将中心元素替换为该中值。这对图像中的椒盐噪声非常有效。有趣的是，在上述过滤器中，中心元素是新计算的值，可能是图像中的像素值或新值。但是在中值模糊中，中心元素总是被图像中的某像素值替换。它有效地降低了噪声。它的内核大小应该是一个正奇数。

在这个演示中，为原始图像添加了 50%的噪点并应用了中值模糊。

```
median = cv.medianBlur(img,5)
```

4. 双边过滤

cv.bilateralFilter()在保持边缘锐利的同时去除噪声非常有效，但与其他过滤器相比，操作速度较慢。如前述可知,高斯滤波器采用像素周围的邻域已找到了其高斯加权平均值。这个高斯滤波器是一个单独的空间函数,即在滤波时考虑附近的像素。它不考虑像素是否具有几乎相同的强度,也不考虑像素是否为边缘像素,所以它也模糊了边缘,但这是不理想的。

双边滤波在空间上也采用了高斯滤波器,但多了一个高斯滤波器,它是像素差的函数。空间的高斯函数确保只考虑对附近的像素进行模糊处理,而强度差异的高斯函数确保只考虑对那些与中心像素具有相似强度的像素进行模糊处理。因为边缘处的像素会有很大的强度变化,所以它保留了边缘。

下面的示例显示了双边滤波器的使用:

```
blur = cv.bilateralFilter(img,9,75,75)
```

1.10　形态变换

形态变换是基于图像形状的一些简单操作。它通常在二进制图像上执行。它需要两个输入:一个是原始图像;另一个被称为结构元素或内核,它决定了操作的性质。两个基本的形态算子是腐蚀和膨胀。然后,它的变体形式(如开运算、闭运算、梯度等)也开始发挥作用。

1. 腐蚀

腐蚀的基本思想就像土壤侵蚀一样,它侵蚀了前景物体的边界(总是尽量保持前景为白色)。那么,它有什么作用呢?内核在图像中滑动(如在2D卷积中)。只有当内核下的所有像素都为1时,原始图像中的某个像素(1或0)才会被认为是1,否则它会被腐蚀(变为0)。

所以发生的事情是,边界附近的所有像素都将根据内核的大小被丢弃。因此,前景物体的厚度或大小会减小,或只是图像中的白色区域减小。它对于去除小的白噪声、分离两个连接的对象等很有用。

在这里,作为一个例子,使用一个5×5的内核。让我们看看它是如何工作的:

```
import cv2 as cv
import numpy as np
img = cv.imread('j.png',0)
kernel = np.ones((5,5),np.uint8)
erosion = cv.erode(img,kernel,iterations = 1)
```

2. 膨胀

膨胀与腐蚀正好相反。这里,如果内核下至少有一个像素为"1",则像素元素为"1"。因此膨胀增加了图像中的白色区域或前景对象的大小。通常,在去除噪声等情况下,腐蚀之后是膨胀。因为腐蚀消除了白噪声,但它也缩小了对象,所以要扩张它。由于噪声消失了,它们不会回来,但对象的面积增加了。膨胀也可用于连接对象的损坏部分。

```
dilation = cv.dilate(img,kernel,iterations = 1)
```

3. 开运算

开运算只是腐蚀后膨胀的另一个名称。它在消除噪声方面很有用。这里我们使用函数cv.morphologyEx():

```
opening = cv.morphologyEx(img, cv.MORPH_OPEN, kernel)
```

4. 闭运算

闭运算与开运算相反，膨胀后腐蚀。它可用于关闭前景对象内部的小孔或对象上的小黑点。

```
closing = cv.morphologyEx(img, cv.MORPH_CLOSE, kernel)
```

5. 形态学梯度

这是用膨胀后的图像减去腐蚀后的图像得到的差值图像，其结果看起来像对象的轮廓。

```
gradient = cv.morphologyEx(img, cv.MORPH_GRADIENT, kernel)
```

6. 礼帽

礼帽是输入图像和进行开运算之后得到的图像的差值图像。下面的示例是针对9×9内核完成的。

```
tophat = cv.morphologyEx(img, cv.MORPH_TOPHAT, kernel)
```

7. 黑帽

黑帽是输入图像和进行闭运算之后得到的图像的差值图像。

```
blackhat = cv.morphologyEx(img, cv.MORPH_BLACKHAT, kernel)
```

在前面的示例中，在 NumPy 的帮助下已手动创建了一个结构元素。它是长方形的，但在某些情况下，可能需要椭圆形或圆形的内核。所以为了这个目的，OpenCV 提供了一个函数 cv.getStructuringElement()。利用这个函数只需传递内核的形状和大小，即可获得所需的内核。示例代码如下：

```
# Rectangular Kernel
>>> cv.getStructuringElement(cv.MORPH_RECT,(5,5))
array([[1, 1, 1, 1, 1],
       [1, 1, 1, 1, 1],
       [1, 1, 1, 1, 1],
       [1, 1, 1, 1, 1],
       [1, 1, 1, 1, 1]], dtype=uint8)
# Elliptical Kernel
>>> cv.getStructuringElement(cv.MORPH_ELLIPSE,(5,5))
array([[0, 0, 1, 0, 0],
       [1, 1, 1, 1, 1],
       [1, 1, 1, 1, 1],
       [1, 1, 1, 1, 1],
       [0, 0, 1, 0, 0]], dtype=uint8)
# Cross-shaped Kernel
>>> cv.getStructuringElement(cv.MORPH_CROSS,(5,5))
array([[0, 0, 1, 0, 0],
       [0, 0, 1, 0, 0],
       [1, 1, 1, 1, 1],
       [0, 0, 1, 0, 0],
       [0, 0, 1, 0, 0]], dtype=uint8)
```

1.11 斑点检测

斑点是图像中的一组连接像素，它们共享一些公共属性，如灰度值。在图 1-1 中，暗连接的区域是斑点,斑点检测的目标是识别和标记这些区域。

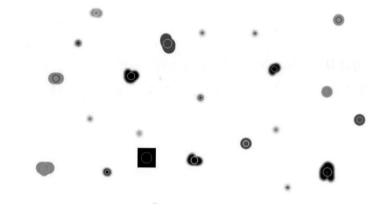

图 1-1　斑点检测

OpenCV 提供了一种方便的方法用来检测斑点,并根据不同的特征对其进行过滤。下面是一个简单的例子:

```python
# Standard imports
import cv2
import numpy as np;

# Read image
im = cv2.imread("blob.jpg", cv2.IMREAD_GRAYSCALE)

# Set up the detector with default parameters
detector = cv2.SimpleBlobDetector()

# Detect blobs.
keypoints = detector.detect(im)

# Draw detected blobs as red circles
# cv2.DRAW_MATCHES_FLAGS_DRAW_RICH_KEYPOINTS ensures the size of the circle corresponds to the size of blob
im_with_keypoints = cv2.drawKeypoints(im, keypoints, np.array([]), (0,0,255), cv2.DRAW_MATCHES_FLAGS_DRAW_RICH_KEYPOINTS)

# Show keypoints
cv2.imshow("Keypoints", im_with_keypoints)
cv2.waitKey(0)
```

顾名思义,SimpleBlobDetector 是基于下面描述的一个相当简单的算法。

1. 算法的步骤

该算法具有以下步骤。

(1) 使用阈值。通过使用从 minThreshold 开始的阈值对源图像进行阈值处理,将源图像转换为多个二值图像。这些阈值按 thresholdStep 递增,直到 maxThreshold。所以,第一个阈值是 minThreshold,第二个是 minThreshold + thresholdStep,第三个是 minThreshold+2 x thresholdStep,依此类推。

(2)分组。在每个二值图像中,连接的白色像素被分组在一起。我们称其为二值斑点。

(3)合并。计算二值图像中二值斑点的中心,并合并比 minDistBetweenBlobs 更近的斑点。

(4)中心和半径计算。计算并返回新合并的斑点的中心和半径。

2. 算法的实现

该算法由参数控制。可以设置 SimpleBlobDetector 的参数来过滤我们想要的斑点类型。

1)按颜色过滤斑点

按颜色过滤斑点,首先,需要设置 filterByColor = 1,设置 blobColor = 0 以选择较暗的斑点,设置 blobColor = 255 则用于选择较亮的斑点。

可以通过设置参数 filterByArea = 1 以及 minArea 和 maxArea 的适当值,根据大小过滤斑点。例如,设置 minArea = 100,将过滤掉所有小于 100 像素的斑点。

2)按形状过滤斑点

现在,形状具有以下三个不同的参数。

(1)圆度。这只是衡量斑点与圆的接近程度。例如。正六边形比正方形具有更高的圆度。要按圆度过滤,需先设置 filterByCircularity = 1,然后为 minCircularity 和 maxCircularity 设置适当的值。圆度定义为 $\frac{4\pi \text{Area}}{(\text{perimeter})^2}$。这意味着圆的圆度为 1,正方形的圆度为 0.785,依此类推。

(2)凸度。凸度定义为 $\frac{\text{斑点的面积}}{\text{凸包的面积}}$。现在,形状的凸包是完全包围该形状的最紧密的凸面形状。为了按凸度过滤,要先设置 filterByConvexity = 1,然后设置 0≤minConvexity≤1 并且 maxConvexity≤1。

(3)惯性比。惯性比可以用来测量形状的拉长程度。例如,对于圆,该值是 1;对于椭圆,它在 0 和 1 之间;对于一条线,它是 0。为了按惯性比过滤,设置 filterByInertia = 1,并适当地设置 0≤minInertiaRatio≤1 和 maxInertiaRatio≤1。

为 SimpleBlobDetector 设置参数很容易。下面是一个例子:

```
# Setup SimpleBlobDetector parameters.
params = cv2.SimpleBlobDetector_Params()

# Change thresholds
params.minThreshold = 10;
params.maxThreshold = 200;

# Filter by Area.
params.filterByArea = True
params.minArea = 1500

# Filter by Circularity
params.filterByCircularity = True
params.minCircularity = 0.1

# Filter by Convexity
params.filterByConvexity = True
```

```
params.minConvexity = 0.87

# Filter by Inertia
params.filterByInertia = True
params.minInertiaRatio = 0.01

# Create a detector with the parameters
ver = (cv2.__version__).split('.')
if int(ver[0]) < 3 :
    detector = cv2.SimpleBlobDetector(params)
else :
    detector = cv2.SimpleBlobDetector_create(params)
```

1.12　Sobel 边缘检测方法

边缘检测需要通过数学方法来查找图像中像素强度的亮度明显变化的点。
- 要做的第一件事是找到灰度图像的梯度,以能够在 x 和 y 方向上找到类似边缘的区域。梯度是导数的多变量推广。虽然可以在单个变量的函数上定义导数,但对于多个变量的函数,梯度会代替它。
- 梯度是向量值函数,与导数相反,导数是标量值。与导数一样,梯度表示函数图形的切线斜率。更准确地说,梯度指向函数增长率最大的方向,其大小是该方向上图形的斜率。

在计算机视觉中,从黑色到白色的过渡被认为一个正斜率,而从白色到黑色的过渡是一个负斜率。

Sobel 边缘检测方法实现的 Python 代码如下:

```
# Python program to Edge detection
# using OpenCV in Python
# using Sobel edge detection
# and laplacian method
import cv2
import numpy as np

#Capture livestream video content from camera 0
cap = cv2.VideoCapture(0)

while(1):

    # Take each frame
    _, frame = cap.read()

    # Convert to HSV for simpler calculations
    hsv = cv2.cvtColor(frame, cv2.COLOR_BGR2HSV)

    # Calculation of Sobelx
    sobelx = cv2.Sobel(frame,cv2.CV_64F,1,0,ksize=5)
```

```
# Calculation of Sobely
sobely = cv2.Sobel(frame,cv2.CV_64F,0,1,ksize=5)

# Calculation of Laplacian
laplacian = cv2.Laplacian(frame,cv2.CV_64F)

cv2.imshow('sobelx',sobelx)
cv2.imshow('sobely',sobely)
cv2.imshow('laplacian',laplacian)
k = cv2.waitKey(5) & 0xFF
if k == 27:
    break

cv2.destroyAllWindows()

#release the frame
cap.release()
```

数字图像由一个矩阵表示。该矩阵在行和列中存储每像素的 RGB/BGR/HSV（图像所属的颜色空间）值。

矩阵的导数由拉普拉斯算子计算。为了计算拉普拉斯算子，需要计算前两个导数，称为 Sobel 导数，每个导数都考虑某个方向的梯度变化：一个是水平方向；另一个是垂直方向。

- 水平索贝尔导数(Sobel x)：通过图像与称为核的矩阵的卷积获得。该矩阵始终具有奇数大小，大小为 3 的内核是最简单的情况。
- 垂直索贝尔导数(Sobel y)：通过图像与称为核的矩阵的卷积获得。该矩阵始终具有奇数大小，大小为 3 的内核是最简单的情况。

1.13 Canny 边缘检测

Canny 边缘检测是一种流行的边缘检测算法。这是一个多阶段算法。

（1）降噪。由于边缘检测容易受到图像中噪声的影响，因此第一步是使用 5×5 高斯滤波器去除图像中的噪声。

（2）寻找图像的强度梯度。计算图像梯度能够得到图像的边缘，因为梯度是灰度变化明显的地方，而边缘也是灰度变化明显的地方。当然，这一步只能得到可能的边缘。因为灰度变化的地方可能是边缘，也可能不是边缘。

（3）非最大抑制。在获得梯度幅度和方向后，对图像进行全面扫描以去除可能不构成边缘的任何不需要的像素。

（4）滞后阈值。这个阶段决定哪些边是真正的边，哪些不是。为此，需要两个阈值 minVal 和 maxVal。任何强度梯度大于 maxVal 的边缘肯定是边缘，低于 minVal 的边缘肯定是非边缘，因此丢弃。位于这两个阈值之间的那些根据它们的连通性被分类为边缘或非边缘。

所以，我们最终得到的是图像中的强边缘。

OpenCV 将上述所有内容放在单个函数 cv.Canny()中。它的第一个参数是输入图像，第二个和第三个参数分别是 minVal 和 maxVal。

```
import numpy as np
import cv2 as cv
from matplotlib import pyplot as plt
img = cv.imread('messi5.jpg',0)
edges = cv.Canny(img,100,200)
plt.subplot(121),plt.imshow(img,cmap = 'gray')
plt.title('Original Image'), plt.xticks([]), plt.yticks([])
plt.subplot(122),plt.imshow(edges,cmap = 'gray')
plt.title('Edge Image'), plt.xticks([]), plt.yticks([])
plt.show()
```

1.14 轮廓检测

使用轮廓检测，可以检测对象的边界，并在图像中轻松定位它们。轮廓检测通常是许多有趣应用的第一步，例如图像前景提取、简单图像分割、检测和识别。

一些非常酷的应用程序使用轮廓进行运动检测或分割。这里有些例子。

- 运动检测。在监控视频中，运动检测技术有很多应用，如室内外安全环境检测、交通控制检测、体育活动中的行为检测、无人看管的物体检测，甚至视频压缩检测。
- 无人看管物体检测。公共场所任何无人看管的物体一般被认为可疑物体。一种有效且安全的解决方案可能是：使用背景减法通过轮廓形成进行无人值守对象检测。
- 背景/前景分割。要将图像的背景替换为另一个，需要执行图像前景提取（类似于图像分割）。使用轮廓是一种可用于执行分割的方法。

当我们连接一个对象边界上的所有点时，我们得到一个轮廓。通常，特定轮廓是指具有相同颜色和强度的边界像素。OpenCV 使在图像中查找和绘制轮廓变得非常容易。它提供了两个简单的函数：findContours()和 drawContours()。

此外，它有两种不同的轮廓检测算法：CHAIN_APPROX_SIMPLE 和 CHAIN_APPROX_NONE。

在 OpenCV 中检测和绘制轮廓的步骤如下：

（1）读取图像并将其转换为灰度格式。

将图像转换为灰度图非常重要，因为它是在为下一步准备图像。将图像转换为单通道灰度图像对于阈值处理很重要，而这又是轮廓检测算法正常工作所必需的。

（2）应用二值阈值。

在寻找轮廓时，首先总是对灰度图像应用二值阈值或 Canny 边缘检测。在这里，我们将应用二值阈值。

这会将图像转换为黑白图，突出显示感兴趣的对象，使轮廓检测算法变得容易。阈值处理使图像中对象的边界完全变白，所有像素都具有相同的强度。该算法现在可以从这些白色像素中检测对象的边界。

注意：值为 0 的黑色像素被视为背景像素并被忽略。

（3）找到轮廓。

使用 findContours()函数检测图像中的轮廓。

（4）在原始 RGB 图像上绘制轮廓。

确定轮廓后，使用 drawContours()函数将轮廓叠加在原始 RGB 图像上。

首先导入 OpenCV，然后读取输入图像。

```
import cv2

# read the image
image = cv2.imread('input/image_1.jpg')
```

接下来，使用 cvtColor()函数将原始 RGB 图像转换为灰度图像。

```
# convert the image to grayscale format
img_gray = cv2.cvtColor(image, cv2.COLOR_BGR2GRAY)
```

现在，使用 threshold()函数将二进制阈值应用于图像。任何值大于 150 的像素都将被设置为值 255（白色）。结果图像中的所有剩余像素都将设置为 0（黑色）。阈值 150 是一个可调参数，因此可以对其进行试验。

阈值化后，使用 imshow()函数可视化二值图像，如下所示。

```
# apply binary thresholding
ret, thresh = cv2.threshold(img_gray, 150, 255, cv2.THRESH_BINARY)
# visualize the binary image
cv2.imshow('Binary image', thresh)
cv2.waitKey(0)
cv2.imwrite('image_thres1.jpg', thresh)
cv2.destroyAllWindows()
```

现在，使用 CHAIN_APPROX_NONE 方法查找并绘制轮廓。

findContours()函数具有如下 3 个必需的参数。

- image：上一步得到的二值输入图像。
- mode：轮廓检索模式。将其作为 RETR_TREE 提供，这意味着该算法将从二值图像中检索所有可能的轮廓。
- method：定义了轮廓近似方法。在这个例子中，将使用 CHAIN_APPROX_NONE。CHAIN_APPROX_NONE 虽然比 CHAIN_APPROX_SIMPLE 稍慢，但我们将在这里使用这个方法来存储所有轮廓点。

这里值得强调的是，mode 是指将要检索的轮廓的类型，而 method 是指存储轮廓内的哪些点。很容易在同一图像上可视化和理解不同方法的结果。因此，在下面的代码示例中，制作原始图像的副本，然后演示方法。

接下来，使用 drawContours()函数在 RGB 图像上叠加轮廓。此函数有 4 个必需的参数和几个可选参数。下面的前 4 个参数是必需的。

- image：要在其上绘制轮廓的输入 RGB 图像。
- contours：表示通过 findContours()函数获得的轮廓。
- contourIdx：得到的轮廓中列出的轮廓点的像素坐标。使用此参数，可以从此列表中指定索引位置，准确指示要绘制的轮廓点。提供负值将绘制所有轮廓点。
- color：要绘制的轮廓点的颜色。下面的 Python 代码中正在绘制绿色点。
- thickness：轮廓点的厚度。

Python 代码如下：

```
# detect the contours on the binary image using cv2.CHAIN_APPROX_NONE
contours, hierarchy = cv2.findContours(image=thresh, mode=cv2.RETR_TREE, method=cv2.CHAIN_APPROX_NONE)
```

```
# draw contours on the original image
image_copy = image.copy()
cv2.drawContours(image=image_copy, contours=contours, contourIdx=-1, color=(0, 255,
0), thickness=2, lineType=cv2.LINE_AA)

# see the results
cv2.imshow('None approximation', image_copy)
cv2.waitKey(0)
cv2.imwrite('contours_none_image1.jpg', image_copy)
cv2.destroyAllWindows()
```

请记住，轮廓算法的准确性和质量在很大程度上取决于提供的二值图像的质量。某些应用需要高质量的轮廓。在这种情况下，在创建二值图像时尝试不同的阈值，看看是否能改善生成的轮廓。

还有其他方法可用于在生成轮廓之前从二值图中消除不需要的轮廓。还可以使用与我们将在此处讨论的轮廓算法相关的更高级功能。

现在让我们找出 CHAIN_APPROX_SIMPLE 算法的工作原理，以及它与 CHAIN_APPROX_NONE 算法的不同之处。

这是 CHAIN_APPROX_SIMPLE 算法的代码：

```
"""
Now let's try with `cv2.CHAIN_APPROX_SIMPLE`
"""
# detect the contours on the binary image using cv2.ChAIN_APPROX_SIMPLE
contours1, hierarchy1 = cv2.findContours(thresh, cv2.RETR_TREE, cv2.CHAIN_APPROX_
SIMPLE)
# draw contours on the original image for `CHAIN_APPROX_SIMPLE`
image_copy1 = image.copy()
cv2.drawContours(image_copy1, contours1, -1, (0, 255, 0), 2, cv2.LINE_AA)
# see the results
cv2.imshow('Simple approximation', image_copy1)
cv2.waitKey(0)
cv2.imwrite('contours_simple_image1.jpg', image_copy1)
cv2.destroyAllWindows()
```

这里唯一的区别是将 findContours() 的方法指定为 CHAIN_APPROX_SIMPLE 而不是 CHAIN_APPROX_NONE。

CHAIN_APPROX_SIMPLE 算法沿轮廓压缩水平、垂直和对角线段，只留下它们的端点。这意味着沿着直线路径的任何点都将被忽略，将只剩下终点。例如，考虑一个沿着矩形的轮廓。除了 4 个角点之外的所有轮廓点都将被消除。这种方法比 CHAIN_APPROX_NONE 更快，因为该算法不存储所有点，使用更少的内存，因此执行时间更少。

层次结构表示轮廓之间的父子关系。您将看到每种轮廓检索模式如何影响图像中的轮廓检测，并产生分层结果。

图像中轮廓检测算法检测到的对象可能是：
- 散布在图像中的单个对象。
- 物体和形状相互交织。

在大多数情况下，当一个形状包含更多形状时，可以安全地得出结论：外部形状是内部形状的父级。

如图 1-10 所示包含几个简单的形状，有助于演示轮廓层次结构。

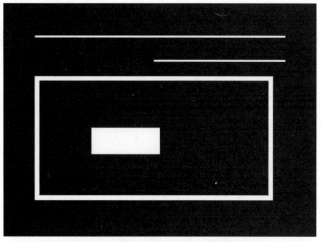

图 1-10　具有简单线条和形状的图像

现在参见图 1-11，其中与图 1-10 中每个形状相关的轮廓已被识别。图 1-11 中的每个数字都具有重要意义。

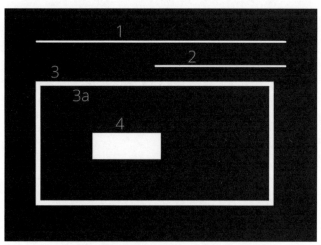

图 1-11　数字显示不同形状之间的父子关系

根据轮廓层次和父子关系，所有单独的数字，即 1、2、3 和 4 都是单独的对象。可以说，3a 是 3 的孩子。注意，3a 代表轮廓 3 的内部部分。

轮廓 1、2 和 4 都是父图形，没有任何关联的子图形，因此它们的编号是任意的。换句话说，轮廓 2 可以被标记为 1，反之亦然。

已经看到 findContours()函数返回两个输出：轮廓列表和层次结构。现在详细讲解轮廓层次结构的输出。

轮廓层次结构表示为一个数组，该数组包含 4 个值，其为[Next, Previous, First_Child, Parent]。那么，所有这些值意味着什么？

Next：表示图像中处于同一层次的下一个轮廓，所以对于轮廓 1，同一层级的下一个轮廓是 2。这里 Next 将是 2。因此，轮廓 3 没有与其自身处于同一层级的轮廓，所以它的 Next 值将是-1。

Previous：表示同一层级的上一个轮廓。这意味着轮廓 1 的 Previous 值始终为-1。

First_Child：表示当前正在考虑的轮廓的第一个子轮廓。轮廓 1 和 2 根本没有孩子，因此它们的 First_Child 的索引值将为-1。但是轮廓 3 有一个孩子，因此对于轮廓 3，First_Child 位置值将是 3a 的索引位置。Parent：表示当前轮廓的父级轮廓的索引位置；很明显，轮廓 1 和 2 没有任何父轮廓；对于轮廓 3a，它的父级将是轮廓 3；对于轮廓 4，父级是轮廓 3a。

上面的解释是有道理的，但如何可视化这些层次数组呢？最好的方法是：

- 使用带有线条和形状的简单图像，如上一个图像。
- 使用不同的检索模式检测轮廓和层次结构。
- 打印值以将它们可视化。

到目前为止，使用了一种特定的检索技术——RETR_TREE 来查找和绘制轮廓，但 OpenCV 还提供了另外 3 种轮廓检索技术，即 RETR_LIST、RETR_EXTERNAL 和 RETR_CCOMP。

所以，现在通过图 1-10 中的图像来查看这 3 种方法，以及与它们相关的代码来获取轮廓。以下代码从磁盘读取图像，将其转换为灰度图，并应用二值化阈值。

```
"""
Contour detection and drawing using different extraction modes to complement
the understanding of hierarchies
"""
image2 = cv2.imread('input/custom_colors.jpg')
img_gray2 = cv2.cvtColor(image2, cv2.COLOR_BGR2GRAY)
ret, thresh2 = cv2.threshold(img_gray2, 150, 255, cv2.THRESH_BINARY)
```

RETR_LIST 轮廓检索方法不会在提取的轮廓之间创建任何父子关系。因此，对于检测到的所有轮廓区域，First_Child 和 Parent 索引位置值始终为-1。如上所述，所有轮廓都将具有其相应的 Previous 和 Next 轮廓。

观察如何在代码中实现 RETR_LIST 方法。

```
contours3, hierarchy3 = cv2.findContours(thresh2, cv2.RETR_LIST, cv2.CHAIN_APPROX_NONE)
image_copy4 = image2.copy()
cv2.drawContours(image_copy4, contours3, -1, (0, 255, 0), 2, cv2.LINE_AA)
# see the results
cv2.imshow('LIST', image_copy4)
print(f"LIST: {hierarchy3}")
cv2.waitKey(0)
cv2.imwrite('contours_retr_list.jpg', image_copy4)
cv2.destroyAllWindows()
```

执行上面的代码会产生以下输出：

```
LIST: [[[ 1 -1 -1 -1]
 [ 2  0 -1 -1]
 [ 3  1 -1 -1]
 [ 4  2 -1 -1]
 [-1  3 -1 -1]]]
```

可以清楚地看到，所有检测到的轮廓区域的第 3 和第 4 索引位置都是-1，正如预期的那样。

RETR_EXTERNAL 轮廓检索方法是一种非常有趣的方法。它只检测父轮廓，而忽略任何子轮廓。因此，像 3a 和 4 这样的所有内部轮廓都不会在其上绘制任何点。

```
contours4, hierarchy4 = cv2.findContours(thresh2, cv2.RETR_EXTERNAL, cv2.CHAIN_APPROX_NONE)
image_copy5 = image2.copy()
cv2.drawContours(image_copy5, contours4, -1, (0, 255, 0), 2, cv2.LINE_AA)
# see the results
cv2.imshow('EXTERNAL', image_copy5)
print(f"EXTERNAL: {hierarchy4}")
cv2.waitKey(0)
cv2.imwrite('contours_retr_external.jpg', image_copy5)
cv2.destroyAllWindows()
```

上面的代码产生以下输出：

```
EXTERNAL: [[[ 1 -1 -1 -1]
[ 2  0 -1 -1]
[-1  1 -1 -1]]]
```

与 RETR_EXTERNAL 不同，RETR_CCOMP 检索图像中的所有轮廓。除此之外，它还对图像中的所有形状或对象应用 2 级层次结构。

这意味着：
- 所有外部轮廓都将具有层次结构级别 1。
- 所有内部轮廓都将具有层次结构级别 2。

但是，如果在另一个层次结构级别为 2 的轮廓内有一个轮廓怎么办？就像在轮廓 3a 之后有轮廓 4 一样。

在这种情况下：
- 同样，轮廓 4 将具有层次结构级别 1。
- 如果轮廓 4 内有任何轮廓，则它们将具有层次结构级别 2。

在图 1-12 中，轮廓已根据其层次级别进行编号，如上所述。

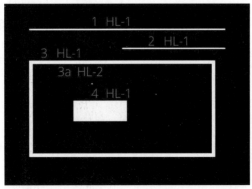

图 1-12　用 RETR_CCOMP 检索方法时显示轮廓中不同层次级别的图像

图 1-12 分别显示了级别 1 和级别 2 的层次结构级别为 HL-1 或 HL-2。现在再来看看代码和输出层次结构数组。

```
contours5, hierarchy5 = cv2.findContours(thresh2, cv2.RETR_CCOMP, cv2.CHAIN_APPROX_NONE)
image_copy6 = image2.copy()
```

```
cv2.drawContours(image_copy6, contours5, -1, (0, 255, 0), 2, cv2.LINE_AA)

# see the results
cv2.imshow('CCOMP', image_copy6)
print(f"CCOMP: {hierarchy5}")
cv2.waitKey(0)
cv2.imwrite('contours_retr_ccomp.jpg', image_copy6)
cv2.destroyAllWindows()
```

执行上面的代码会产生以下输出：

```
CCOMP: [[[ 1 -1 -1 -1]
 [ 3  0  2 -1]
 [-1 -1 -1  1]
 [ 4  1 -1 -1]
 [-1  3 -1 -1]]]
```

就像 RETR_CCOMP 一样，RETR_TREE 也检索所有轮廓。它还创建了一个级别不限于 1 或 2 的完整的层次结构。每个轮廓可以有自己的层次结构，如图 1-13 所示。

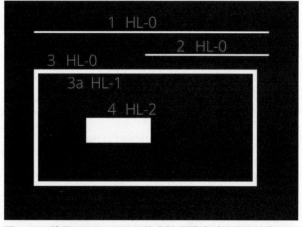

图 1-13 使用 RETR_TREE 轮廓检索模式时的层次结构级别

从图 1-13 可以看出：
- 轮廓 1、2 和 3 处于同一级别，即级别 0。
- 轮廓 3a 存在于层次结构级别 1，因为它是轮廓 3 的子级。
- 轮廓 4 是一个新的轮廓区域，所以它的层级为 2。

以下代码使用 RETR_TREE 模式检索轮廓。

```
contours6, hierarchy6 = cv2.findContours(thresh2, cv2.RETR_TREE, cv2.CHAIN_APPROX_NONE)
image_copy7 = image2.copy()
cv2.drawContours(image_copy7, contours6, -1, (0, 255, 0), 2, cv2.LINE_AA)
# see the results
cv2.imshow('TREE', image_copy7)
print(f"TREE: {hierarchy6}")
cv2.waitKey(0)
cv2.imwrite('contours_retr_tree.jpg', image_copy7)
cv2.destroyAllWindows()
```

执行上面的代码会产生以下输出：

```
TREE: [[[ 3 -1  1 -1]
  [-1 -1  2  0]
  [-1 -1 -1  1]
  [ 4  0 -1 -1]
  [-1  3 -1 -1]]]
```

当图像中对象的背景充满线条时,轮廓算法可能无法提供有意义且有用的结果。

1.15 人脸检测

本节将讲解如何使用 OpenCV 在 Python 中构建一个简单的人脸检测器。下面编写一个程序来检测图像、视频中的人脸或直接从相机中检测人脸。

首先,安装包,打开 Python 终端并输入命令。

```
pip install opencv-python
pip install imutils
```

安装完成后,就可以将其导入程序了。

```
import cv2
import imutils
```

使用的 Haar Cascades 人脸检测是一种机器学习方法,其中使用一组输入数据训练级联函数。OpenCV 已经包含许多针对面部、眼睛、微笑等的预训练分类器……下面将使用面部分类器从图像、视频或相机中检测面部。为此,需要下载 Haar Cascade 的 XML 文件。该文件可以从 https://raw.githubusercontent.com/opencv/opencv/master/data/haarcascades/haarcascade_frontalface_default.xml 链接下载。将文件保存在当前文件夹后,将其加载到程序中。

```
# Load the cascade
face_cascade = cv2.CascadeClassifier('haarcascade_frontalface_default.xml')
```

在这一步,将选择一幅用于测试代码的图像,且确保图像中至少有一张脸,以便程序可以找到一张脸。

选择图像后,即可在程序中定义它,并确保图像文件位于正在使用的同一文件夹中。

```
# Read the input image
img = cv2.imread('test.jpg')
img_resize = imutils.resize(img, width = 800)
gray = cv2.cvtColor(img_resize, cv2.COLOR_BGR2GRAY)
```

如下是检测图像中人脸的代码:

```
# Detect Faces
faces = face_cascade.detectMultiScale(gray, 1.3, 5)
```

detectMultiScale()用于检测人脸。它需要以下 3 个参数。

- 输入图像。
- scaleFactor 指定图像大小按比例缩放多少,最佳值为 1.3。
- minNeighbours 指定构成检测目标的相邻矩形的最小个数,它的最佳值是 5。

之前在代码中已定义了 face_cascade。检测到人脸后,程序将在人脸的周围绘制矩形,以示机器检测到了什么。机器可能会犯错误,但程序的目标是更优化和准确。

要在检测到的人脸周围绘制一个矩形,必须编写以下代码:

```
# Draw rectangle around the face
for (x, y, w, h) in faces:
    cv2.rectangle(img_resize, (x,y), (x+w, y+h), (255, 0, 0), 2)
```

这里的参数（255,0,0）表示矩形的颜色。

参数 2 表示线的粗细。

下面的代码将结果显示为一个弹出窗口：

```
# Export the result
cv2.imshow('face_detected', img_resize)
cv2.waitKey(0)
cv2.destroyAllWindows()
cv2.waitKey(1)
```

导入 OpenCV 库（cv2）以后，从网络摄像头捕获视频或输入视频文件。

```
# Read the video file
# video = cv2.VideoCapture('video-3.mp4')

# Open Webcam
video = cv2.VideoCapture(0)
```

众所周知，视频是使用帧创建的，而帧是静止图像。所以，我们将对视频中的每帧进行人脸检测。

```
while True:
    ret,frame = video.read()
    video_gray = cv2.cvtColor(frame, cv2.COLOR_BGR2GRAY)
    # Detect Faces
    faces = face_cascade.detectMultiScale(video_gray, 1.3, 4)
    # Draw triangle around the face
    for (x, y, w, h) in faces:
        cv2.rectangle(frame, (x,y), (x+w, y+h), (255, 0, 0), 2)

    # Show frame
    cv2.imshow('Face Detected', frame)
    k = cv2.waitKey(30)
    if k == 27:
        break

video.release()
cv2.destroyAllWindows()
```

1.16 特征脸

本节将讲解特征脸——一种用于人脸的主成分分析（PCA）应用。

特征脸是可以添加到平均人脸以创建一张新人脸图像的图像。其在数学表达式为

$$F = F_m + \sum_{i=1}^{n} \alpha_i F_i$$

式中，F 为一张新脸；F_m 为平均人脸；F_i 为一张特征脸；α_i 为可以选择创建新脸的标量乘

数,可以是正数或负数。

通过估计面部图像数据集的主成分来计算特征脸,可应用于人脸识别和面部地标检测等。计算特征脸需要以下步骤。

(1)获取面部图像数据集。需要包含不同类型面部的面部图像集合。这里,使用了来自 CelebA(http://mmlab.ie.cuhk.edu.hk/projects/CelebA.html)的大约 200 幅图像。

(2)对齐和调整图像大小。需要对齐和调整图像大小,使眼睛的中心在所有图像中对齐。这可以通过首先找到面部标志来完成。这里,使用了 CelebA 提供的对齐图像。此时,数据集中的所有图像应该是大小相同的。

(3)创建数据矩阵。创建一个包含所有图像作为行向量的数据矩阵。如果数据集中所有图像的大小都为 100×100 像素并且有 1000 幅图像,那么将有一个大小为 30000×1000 个元素的数据矩阵。

(4)计算平均向量(可选)。在对数据进行 PCA 之前,需要减去平均向量。例如,平均向量是一个 30000×1 个元素的行向量,通过平均数据矩阵的所有行来计算。使用 OpenCV 的 PCA 类不需要计算这个均值向量的原因是,如果没有提供向量,OpenCV 也可以方便地计算出均值。在其他线性代数包中可能不是这种情况。

(5)计算主成分。该数据矩阵的主成分是通过查找协方差矩阵的特征向量来计算的。幸运的是,OpenCV 中的 PCA 类处理这个计算,只需要提供数据矩阵,就可以得到一个包含特征向量的矩阵。

(6)重塑特征向量以获得特征脸。如果数据集包含向量长度为 100×100×3 的图像,则如此获得的特征向量的长度为 30000 个元素。因此,可以将这些特征向量重塑为 100×100×3 的图像以获得特征脸。

OpenCV 中的 PCA 类允许计算数据矩阵的主成分。这里,正在讨论的是使用 PCA 类的最常见方法。

```
mean, eigenvectors = cv2.PCACompute ( data, mean=mean, maxComponents=maxComponents )
```

这里,
- data:包含作为行或列向量的每个数据点的数据矩阵。如果数据包含 1000 幅图像,并且每幅图像是一个 30000 个元素的行向量,那么数据矩阵的大小将是 30000×1000 个元素。
- mean:数据的平均值。如果数据矩阵中的每个数据点都是一个 30000 个元素的行向量,则平均值也将是一个相同大小的向量。此参数是可选的,如果未提供,则在内部计算。
- maxComponents:主成分的最大数量通常是两个值中较小的一个:原始数据的维度(如 30000);数据点的数量(如 1000 个)。但是,可以通过设置这个参数来明确地确定要计算的最大组件数。例如,可能只对前 50 个主成分感兴趣。

让我们回顾一下 Python 中的 main()函数。

```
if __name__ == '__main__':

    # Number of EigenFaces
    NUM_EIGEN_FACES = 10

    # Maximum weight
    MAX_SLIDER_VALUE = 255
```

```python
# Directory containing images
dirName = "images"

# Read images
images = readImages(dirName)

# Size of images
sz = images[0].shape

# Create data matrix for PCA.
data = createDataMatrix(images)

# Compute the eigenvectors from the stack of images created
print("Calculating PCA ", end="...")
mean, eigenVectors = cv2.PCACompute(data, mean=None, maxComponents=NUM_EIGEN_FACES)
print ("DONE")

averageFace = mean.reshape(sz)

eigenFaces = [];

for eigenVector in eigenVectors:
    eigenFace = eigenVector.reshape(sz)
    eigenFaces.append(eigenFace)

# Create window for displaying Mean Face
cv2.namedWindow("Result", cv2.WINDOW_AUTOSIZE)

# Display result at 2x size
output = cv2.resize(averageFace, (0,0), fx=2, fy=2)
cv2.imshow("Result", output)

# Create Window for trackbars
cv2.namedWindow("Trackbars", cv2.WINDOW_AUTOSIZE)

sliderValues = []

# Create Trackbars
for i in xrange(0, NUM_EIGEN_FACES):
    sliderValues.append(MAX_SLIDER_VALUE/2)
    cv2.createTrackbar( "Weight" + str(i), "Trackbars", MAX_SLIDER_VALUE/2, MAX_SLIDER_VALUE, createNewFace)

# You can reset the sliders by clicking on the mean image
cv2.setMouseCallback("Result", resetSliderValues);

print('''Usage:
Change the weights using the sliders
Click on the result window to reset sliders
Hit ESC to terminate program.''')
```

```
cv2.waitKey(0)
cv2.destroyAllWindows()
```

上面的代码执行以下操作。

（1）将特征脸数（NUM_EIGEN_FACES）设置为 10，将滑块的最大值（MAX_SLIDER_VALUE）设置为 255。这些数字并非一成不变。可以更改这些数字以查看应用程序如何更改。

（2）读取图像。使用函数 readImages()读取指定目录中的所有图像。该目录包含对齐的图像，而且所有图像中左右眼的中心是相同的。可以将这些图像添加到列表（或向量）中，还可以垂直翻转图像并将它们添加到列表中。因为有效面部图像的镜像，只是将数据集的大小增加了一倍并同时使其对称。

（3）组装数据矩阵。使用函数 createDataMatrix()将图像组装成数据矩阵。数据矩阵的每行是一幅图像。下面来看看 createDataMatrix()函数：

```
def createDataMatrix(images):
    print("Creating data matrix",end="... ")
    '''
    Allocate space for all images in one data matrix.
    The size of the data matrix is
    ( w  * h  * 3, numImages )
    where,
    w = width of an image in the dataset.
    h = height of an image in the dataset.
    3 is for the 3 color channels.
    '''

    numImages = len(images)
    sz = images[0].shape
    data = np.zeros((numImages, sz[0] * sz[1] * sz[2]), dtype=np.float32)
    for i in range(0, numImages):
        image = images[i].flatten()
        data[i,:] = image

    print("DONE")
    return data
```

（4）计算 PCA。使用 Python 中的 PCACompute()函数来计算 PCA。作为 PCA 的输出，获得了平均向量和 10 个特征向量。

（5）重塑向量以获得平均人脸和特征脸。平均向量和每个特征向量的长度为 $w×h×3$，其中 w 是宽度；h 是高度；3 是数据集中任何图像的颜色通道数。换句话说，它们是具有 30000 个元素的向量。将它们重塑为图像的原始大小，以获得平均人脸和特征脸。

（6）根据滑块值创建新脸。可以通过使用函数 createNewFace()将加权特征脸添加到平均人脸来创建新人脸。在 OpenCV 中，滑块值不能为负。所以可以通过从当前滑块值中减去 MAX_SLIDER_VALUE/2 来计算权重，这样就可以获得正值和负值。

```
def createNewFace(*args):
    # Start with the mean image
    output = averageFace
```

```
    # Add the eigen faces with the weights
    for i in range(0, NUM_EIGEN_FACES):
        '''
        OpenCV does not allow slider values to be negative.
        So we use weight = sliderValue - MAX_SLIDER_VALUE / 2
        '''
        sliderValues[i] = cv2.getTrackbarPos("Weight" + str(i), "Trackbars");
        weight = sliderValues[i] - MAX_SLIDER_VALUE/2
        output = np.add(output, eigenFaces[i] * weight)

    # Display Result at 2x size
    output = cv2.resize(output, (0,0), fx=2, fy=2)
    cv2.imshow("Result", output)
```

1.17 图像金字塔

在进行图像处理时，通常使用默认分辨率的图像，但很多时候需要更改分辨率（降低它）或调整原始图像的大小，在这种情况下，图像金字塔会派上用场。

pyrUp()函数将图像大小增大到其原始大小的 2 倍，而 pyrDown()函数将图像大小减小到其原始大小的一半。如果将原始图像保留为基础图像并继续对其应用 pyrDown()函数并将图像保持在垂直堆栈中，它将看起来像一个金字塔。通过 pyrUp()函数对原始图像进行放大也是如此。

一旦图像被缩小，如果再将其重新缩放到原始大小，则会丢失一些信息，并且新图像的分辨率比原始图像的分辨率低得多。

下面是一个图像金字塔的例子：

```
import cv2
import matplotlib.pyplot as plt

img = cv2.imread("images/input.jpg")

layer = img.copy()

for i in range(4):
    plt.subplot(2, 2, i + 1)

    # using pyrDown() function
    layer = cv2.pyrDown(layer)

    plt.imshow(layer)
    cv2.imshow("str(i)", layer)
    cv2.waitKey(0)


cv2.destroyAllWindows()
```

1.18 实时人体检测

下面使用 OpenCV 中实现的 HOG 算法来实时检测视频流中的人物。

使用 OpenCV 从网络摄像头读取数据很容易，只需编写以下脚本并使用 Python 运行它：

```
import numpy as np
import cv2

cv2.startWindowThread()
cap = cv2.VideoCapture(0)

while(True):
    # reading the frame
    ret, frame = cap.read()
    # displaying the frame
    cv2.imshow('frame',frame)
    if cv2.waitKey(1) & 0xFF == ord('q'):
        # breaking the loop if the user types q
        # note that the video window must be highlighted!
        break

cap.release()
cv2.destroyAllWindows()
# the following is necessary on the mac,
# maybe not on other platforms:
cv2.waitKey(1)
```

应该会看到一个弹出窗口，其中包含来自摄像头的图像。

现在让我们尝试操作视频流。视频是逐帧读取的，因此可以在显示之前对其进行编辑。在显示帧之前添加以下代码：

```
# turn to greyscale:
frame = cv2.cvtColor(frame, cv2.COLOR_RGB2GRAY)
# apply threshold. all pixels with a level larger than 80 are shown in white. the others
are shown in black:
ret,frame = cv2.threshold(frame,80,255,cv2.THRESH_BINARY)
```

OpenCV 实现了一种非常快速的人体检测方法，称为 HOG（定向梯度直方图）。

这种方法被用来训练检测行人，这些行人大多是站着的，并且完全可见。不要指望它在其他情况下能很好地工作。

在讨论这个方法之前，先修改脚本如下：

```
# import the necessary packages
import numpy as np
import cv2


# initialize the HOG descriptor/person detector
hog = cv2.HOGDescriptor()
hog.setSVMDetector(cv2.HOGDescriptor_getDefaultPeopleDetector())
```

```python
cv2.startWindowThread()

# open webcam video stream
cap = cv2.VideoCapture(0)

# the output will be written to output.avi
out = cv2.VideoWriter(
    'output.avi',
    cv2.VideoWriter_fourcc(*'MJPG'),
    15.,
    (640,480))

while(True):
    # Capture frame-by-frame
    ret, frame = cap.read()

    # resizing for faster detection
    frame = cv2.resize(frame, (640, 480))
    # using a greyscale picture, also for faster detection
    gray = cv2.cvtColor(frame, cv2.COLOR_RGB2GRAY)

    # detect people in the image
    # returns the bounding boxes for the detected objects
    boxes, weights = hog.detectMultiScale(frame, winStride=(8,8) )

    boxes = np.array([[x, y, x + w, y + h] for (x, y, w, h) in boxes])

    for (xA, yA, xB, yB) in boxes:
        # display the detected boxes in the colour picture
        cv2.rectangle(frame, (xA, yA), (xB, yB),
                      (0, 255, 0), 2)

    # Write the output video
    out.write(frame.astype('uint8'))
    # Display the resulting frame
    cv2.imshow('frame',frame)
    if cv2.waitKey(1) & 0xFF == ord('q'):
        break

# When everything done, release the capture
cap.release()
# and release the output
out.release()
# finally, close the window
cv2.destroyAllWindows()
cv2.waitKey(1)
```

现在运行脚本会发现，如果此人离相机不太近，则检测器的效果会更好；如果这个人离相机很近，通常会显示几个重叠的框。

1.19 背景减除

捕获视频用于背景减除。

```
cap = cv.VideoCapture('./img/vtest.avi')
```

VideoCapture()是一个 OpenCV 函数,它需要一个参数,即视频的来源。0 表示前置摄像头,1 表示后置摄像头,对于录制的视频,提供绝对或相对路径。

OpenCV 中有两个用于背景减除的函数,即 MOG2 和 KNN。

```
fgbg = cv.createBackgroundSubtractorMOG2(detectShadows=False)
```

如果不想检测阴影并将其从图像中删除,请将参数 detectShadows 传递为 False。

KNN 背景减除代码如下:

```
fgbg = cv.createBackgroundSubtractorKNN(detectShadows=False)
```

这是另一种背景减除算法,称为 KNN。它还将检测阴影参数视为 True 或 False。这个 OpenCV 函数将初始化背景减除。

读帧:

```
ret, frame = cap.read()
```

在 OpenCV 中应用背景减除:

```
fgmask = fgbg.apply(frame)
```

在 MOG2 和 KNN 背景减除方法/步骤中,已创建了一个背景减除实例,并将该实例命名为 fgbg。

现在,在视频的每帧中使用 apply()函数来去除背景。apply()函数将必须从中删除背景的源图像/帧作为参数。

然后使用 cv.imshow()函数显示去除背景的图像"fgmask"。

减去的帧使用 while 循环显示,当程序能够读取视频时该循环为真。所以需要一些逻辑来结束程序,否则程序将陷入无限循环。

结束无限循环的一种方法是将键盘键绑定到将中断循环的程序。

在这种情况下,正在绑定键盘键'q'来结束循环,之后所有创建的窗口将被销毁并释放所占用的内存。

在 OpenCV 中创建的背景减除函数的完整代码如下。

```python
def bs():
    cap = cv.VideoCapture('./img/vtest.avi')
    
    # fgbg = cv.createBackgroundSubtractorMOG2(detectShadows=False)
    fgbg = cv.createBackgroundSubtractorKNN(detectShadows=False)
    
    while cap.isOpened():
        ret, frame = cap.read()
        
        if frame is None:
            break
        
        fgmask = fgbg.apply(frame)
```

```
        cv.imshow('Frame', frame)
        cv.imshow('FG Mask', fgmask)
        if cv.waitKey(30) & 0xFF == ord('q'):
            break

cap.release()
cv.destroyAllWindows()
```

1.20 模板匹配

模板匹配是一种在较大图像中搜索和查找模板图像位置的方法。为此，OpenCV 附带了一个函数 cv.matchTemplate()。它只是将模板图像滑到输入图像上，并在模板图像下比较输入图像的模板和补丁。OpenCV 中实现了几种比较方法。它返回一个灰度图像，其中每像素表示该像素的邻域与模板匹配的程度。

如果输入图像的大小为（W×H），模板图像的大小为（w×h），则输出图像的大小为（W-w+1，H-h+1）。得到结果后，可以使用 cv.minMaxLoc()函数查找最大值/最小值在哪里。把它作为矩形的左上角，把（w, h）作为矩形的宽和高。该矩形就是模板区域。

这里，作为一个例子，通过在梅西的照片中搜索梅西的脸来尝试所有的比较方法，以便可以看到它们的结果如何：

```
import cv2 as cv
import numpy as np
from matplotlib import pyplot as plt
img = cv.imread('messi5.jpg',0)
img2 = img.copy()
template = cv.imread('template.jpg',0)
w, h = template.shape[::-1]
# All the 6 methods for comparison in a list
methods = ['cv.TM_CCOEFF', 'cv.TM_CCOEFF_NORMED', 'cv.TM_CCORR',
            'cv.TM_CCORR_NORMED', 'cv.TM_SQDIFF', 'cv.TM_SQDIFF_NORMED']
for meth in methods:
    img = img2.copy()
    method = eval(meth)
    # Apply template Matching
    res = cv.matchTemplate(img,template,method)
    min_val, max_val, min_loc, max_loc = cv.minMaxLoc(res)
    # If the method is TM_SQDIFF or TM_SQDIFF_NORMED, take minimum
    if method in [cv.TM_SQDIFF, cv.TM_SQDIFF_NORMED]:
        top_left = min_loc
    else:
        top_left = max_loc
    bottom_right = (top_left[0] + w, top_left[1] + h)
    cv.rectangle(img,top_left, bottom_right, 255, 2)
    plt.subplot(121),plt.imshow(res,cmap = 'gray')
    plt.title('Matching Result'), plt.xticks([]), plt.yticks([])
```

```
    plt.subplot(122),plt.imshow(img,cmap = 'gray')
    plt.title('Detected Point'), plt.xticks([]), plt.yticks([])
    plt.suptitle(meth)
plt.show()
```

在上个例子中,在图像中搜索了梅西的脸,这在图像中只出现了一次。假设正在搜索一个多次出现的对象,cv.minMaxLoc()不会提供所有位置。在这种情况下,可以使用阈值。所以在这个例子中,将使用著名游戏《马里奥》中的截图,并在其中找硬币。

```
import cv2 as cv
import numpy as np
from matplotlib import pyplot as plt
img_rgb = cv.imread('mario.png')
img_gray = cv.cvtColor(img_rgb, cv.COLOR_BGR2GRAY)
template = cv.imread('mario_coin.png',0)
w, h = template.shape[::-1]
res = cv.matchTemplate(img_gray,template,cv.TM_CCOEFF_NORMED)
threshold = 0.8
loc = np.where( res >= threshold)
for pt in zip(*loc[::-1]):
    cv.rectangle(img_rgb, pt, (pt[0] + w, pt[1] + h), (0,0,255), 2)
cv.imwrite('res.png',img_rgb)
```

1.21 直线检测

霍夫变换是一种在图像处理中用于检测任何形状的方法,如果该形状可以用数学形式表示,即使形状有点破损或变形,它也可以检测到。

下面介绍霍夫变换如何使用 HoughLine 变换方法进行直线检测。要应用 HoughLine 方法,首先需要对特定图像进行边缘检测。

一条线可以表示为 $r = x\cos\theta + y\sin\theta$。其中,$r$ 是从原点到直线的垂直距离;θ 是该垂直线与水平轴形成的角度。

为什么不使用下面给出的人们熟悉的方程

$$y = mx + c$$

原因是斜率 m 可以取 $-\infty \sim +\infty$ 的值。而对于霍夫变换,参数需要有界。

可能还有一个后续问题:在 (r, θ) 形式中,θ 是有界的,但 r 不能取 $0 \sim +\infty$ 的值吗?这在理论上可能是正确的,但在实践中,r 也是有界的,因为图像本身是有限的。

在 OpenCV 中,使用霍夫变换的直线检测是在函数 HoughLines 和 HoughLinesP(概率霍夫变换)中实现的。此函数采用以下参数:

- edges:边缘检测器的输出。
- lines:用于存储线的起点和终点坐标的向量。
- rho:分辨率参数 r 以像素为单位。
- theta:以弧度为单位的参数 θ 的分辨率。
- threshold:检测直线的最小交叉点数。

检测直线的 Python 代码如下:

```
# Read image
img = cv2.imread('lanes.jpg', cv2.IMREAD_COLOR) # road.png is the filename
# Convert the image to gray-scale
gray = cv2.cvtColor(img, cv2.COLOR_BGR2GRAY)
# Find the edges in the image using canny detector
edges = cv2.Canny(gray, 50, 200)
# Detect points that form a line
lines = cv2.HoughLinesP(edges, 1, np.pi/180, max_slider, minLineLength=10, maxLineGap=250)
# Draw lines on the image
for line in lines:
    x1, y1, x2, y2 = line[0]
    cv2.line(img, (x1, y1), (x2, y2), (255, 0, 0), 3)
# Show result
cv2.imshow("Result Image", img)
```

1.22 霍夫圆变换

一个圆在数学上表示为$(x-x_{center})^2+(y-y_{center})^2=r^2$，其中$(x_{center}, y_{center})$是圆心，$r$是圆的半径。从方程中，可以看到圆有 3 个参数，所以需要一个 3D 累加器来进行霍夫变换，但这将是非常无效的。因此，OpenCV 使用了更复杂的方法，即霍夫梯度法，它使用边缘的梯度信息。

这里使用的函数是 cv.HoughCircles()。该函数的参数如下：

（1）image：8 位、单通道、灰度输入图像。

（2）method：检测方法。可用的方法是 HOUGH_GRADIENT 和 HOUGH_GRADIENT_ALT。

（3）dp：累加器分辨率与图像分辨率的反比。例如，如果 dp=1，则累加器具有与输入图像相同的分辨率。如果 dp=2，累加器的宽度和高度只有一半。对于 HOUGH_GRADIENT_ALT 方法，推荐值为 dp=1.5，除非需要检测一些非常小的圆圈。

（4）minDist：检测到的圆的中心之间的最小距离。如果参数太小，除了一个真实的圆圈外，可能还会错误地检测到多个相邻圆圈；如果参数太大，可能会漏掉一些圆圈。

（5）param1：第一个特定于方法的参数。在 HOUGH_GRADIENT 和 HOUGH_GRADIENT_ALT 的情况下，它是传递给 Canny 边缘检测器的高阈值（常取低阈值的 2 倍）。

（6）param2：第二个特定于方法的参数。在 HOUGH_GRADIENT 方法下，它是检测阶段圆心的累加器阈值。它越小，可能检测到的错误圆圈就越多。与较大的累加器值相对应的圆圈将首先返回。在 HOUGH_GRADIENT_ALT 方法下，这是圆形"完美度"度量。它越接近 1，形状圆算法选择得越好。在大多数情况下，0.9 应该没问题。如果想更好地检测小圆圈，可以将其降低到 0.85、0.8，甚至更低。但是接下来也要尽量限制搜索范围[minRadius, max Radius]，避免出现很多假圈。

（7）minRadius：最小圆半径。

（8）maxRadius：最大圆半径。如果≤0，则使用最大图像尺寸。

查找圆的代码如下：

```
import numpy as np
import cv2 as cv
img = cv.imread('opencv-logo-white.png',0)
```

```
img = cv.medianBlur(img,5)
cimg = cv.cvtColor(img,cv.COLOR_GRAY2BGR)
circles = cv.HoughCircles(img,cv.HOUGH_GRADIENT,1,20,
                            param1=50,param2=30,minRadius=0,maxRadius=0)
circles = np.uint16(np.around(circles))
for i in circles[0,:]:
    # draw the outer circle
    cv.circle(cimg,(i[0],i[1]),i[2],(0,255,0),2)
    # draw the center of the circle
    cv.circle(cimg,(i[0],i[1]),2,(0,0,255),3)
cv.imshow('detected circles',cimg)
cv.waitKey(0)
cv.destroyAllWindows()
```

1.23 镜头畸变

为了生成清晰锐利的图像，针孔相机光圈的直径应尽可能小。如果增加光圈的大小，则来自物体多个点的光线会入射到屏幕的同一部分，从而产生模糊的图像。

一方面，如果将光圈尺寸缩小，则只有少量光子撞击图像传感器。结果，图像又暗又嘈杂。因此，针孔相机的光圈越小，图像越聚焦，但同时也越暗，噪点越多。

另一方面，光圈越大，图像传感器接收的光子就越多（因此信号也更多）。这导致只有少量噪声的明亮图像。

如何获得清晰的图像，同时捕获更多的光线以使图像明亮？

可以用透镜代替针孔，从而增加光线可以通过的光圈大小。镜头允许大量光线通过孔，并且由于其光学特性，它还可以将它们聚焦在屏幕上。这使图像更亮。

因此，使用镜头获得了明亮、锐利、聚焦的图像。

问题已经解决了？没那么快。没有什么是免费的！

通过使用镜头，虽然可以获得更好质量的图像，但是镜头会引入一些失真效果。失真效果主要有以下两种类型。

（1）径向畸变：这种类型的畸变通常是由于光线的不均匀弯曲而发生的。与靠近透镜中心的光线相比，靠近镜片边缘的光线弯曲得更多。由于径向畸变，现实世界中的直线在图像中似乎是弯曲的。光线在撞击图像传感器之前从其理想位置径向向内或向外移动。有两种径向畸变效果：

①桶形畸变效应，对应于负径向位移。

②枕形畸变效应，对应于正径向位移。

（2）切向畸变：这通常发生在图像屏幕或传感器与镜头成一定角度时。因此，图像似乎是倾斜和拉伸的。

根据畸变的来源，畸变有 3 种类型，即径向畸变、偏心畸变和薄棱镜畸变。偏心和薄棱镜畸变同时具有径向和切向畸变效果。

消除因镜头引起的畸变需要以下 3 个主要步骤。

（1）执行相机标定并获取相机的内在参数。

（2）优化相机矩阵以控制未失真图像中不需要的像素的百分比。

(3)使用细化的相机矩阵使图像不失真。

消除畸变的代码如下:

```
# Refining the camera matrix using parameters obtained by calibration
newcameramtx, roi = cv2.getOptimalNewCameraMatrix(mtx, dist, (w,h), 1, (w,h))

# Method 1 to undistort the image
dst = cv2.undistort(img, mtx, dist, None, newcameramtx)

# Method 2 to undistort the image
mapx,mapy=cv2.initUndistortRectifyMap(mtx,dist,None,newcameramtx,(w,h),5)

dst = cv2.remap(img,mapx,mapy,cv2.INTER_LINEAR)

# Displaying the undistorted image
cv2.imshow("undistorted image",dst)
cv2.waitKey(0)
```

1.24 使用 Hu 矩进行形状匹配

本节将介绍如何使用 Hu 矩进行形状匹配。

图像矩是图像像素强度的加权平均值。为简单起见,考虑单通道二值图像 I。位置(x, y)处的像素强度由 $I(x, y)$ 给出。注意,对于二值图像 $I(x, y)$ 可以取值 0 或 1。

可以定义最简单的矩为

$$M=\sum_x \sum_y I(x,y)$$

在上述等式中所做的只是计算所有像素强度的总和。换句话说,所有像素的强度仅根据它们的强度进行加权,而不是基于它们在图像中的位置。

对于二值图像,上述矩可以用以下几种不同的方式解释:

(1)它是白色像素的数量(即强度 = 1)。
(2)它是图像中白色区域的面积。

对于两个相同的形状,上述图像矩必然相同,但这不是充分条件。可以很容易地构造两个图像,上面的矩是相同的,但它们看起来非常不同。

下面是一些更复杂的矩。

$$M_{ij}=\sum_x \sum_y x^i y^j I(x,y)$$

式中,i 和 j 是整数(如 0, 1, 2, …)。$I(x, y)$ 表示位置(x, y)处的像素强度。这些矩通常被称为原始矩,以将它们与本文后面提到的中心矩区分开来。

请注意,上述矩取决于像素的强度和它们在图像中的位置。所以这些矩捕捉了一些形状的信息。

质心($\bar{x}, \bar{y}$)使用以下公式计算。

$$\bar{x} = \frac{M_{10}}{M_{00}}$$

$$\bar{y} = \frac{M_{01}}{M_{00}}$$

中心矩之前所介绍的原始图像矩非常相似，只是在矩公式中从 x 和 y 中减去了质心。

$$\mu_{ij} = \sum_x \sum_y (x-\bar{x})^i (y-\bar{y})^j I(x,y)$$

请注意，上述中心矩是平移不变的。换句话说，无论斑点在图像中的哪个位置，如果形状相同，那么矩都是相同的。

如果也可以使矩保持尺度不变，那不是很酷吗？好吧，为此需要归一化的中心矩：

$$\eta_{ij} = \frac{\mu_{i,j}}{\mu_{00}^{(i+j)/2+1}}$$

中心矩是平移不变的，这很好。但这对于形状匹配是不够的。理想的是计算对平移、缩放和旋转不变的矩。

幸运的是，实际上可以计算出这样的矩，它们被称为 Hu 矩。

在 OpenCV 中，使用 HuMoments() 来计算输入图像中存在的形状的 Hu 矩。

首先，将图像读取为灰度图像。

```
# Read image as grayscale image
im = cv2.imread(filename,cv2.IMREAD_GRAYSCALE)
```

使用阈值对图像进行二值化：

```
# Threshold image
_,im = cv2.threshold(im, 128, 255, cv2.THRESH_BINARY)
```

OpenCV 有一个用于计算 Hu 矩的内置函数。毫不奇怪，它被称为 HuMoments。它将图像的中心矩作为输入，可以使用函数 moments() 计算中心矩：

```
# Calculate Moments
moments = cv2.moments(im)
# Calculate Hu Moments
huMoments = cv2.HuMoments(moments)
```

上一步得到的 Hu 矩范围很大，但可以使用下面给出的对数变换将它们置于相同的范围内。

$$H_i = -\text{sign}(h_i) \log |h_i|$$

经过上述变换后，矩具有可比性。对数比例变换的代码如下：

```
# Log scale hu moments
for i in range(0,7):
    huMoments[i] = -1* copysign(1.0, huMoments[i]) * log10(abs(huMoments[i])))
```

OpenCV 提供了一个易于使用的实用函数，称为 matchShapes()，它接收两个图像（或轮廓）并使用 Hu 矩找到它们之间的距离。因此，不必显式计算 Hu 矩，而只须将图像二值化并使用函数 matchShapes()。其用法如下：

```
d1 = cv2.matchShapes(im1,im2,cv2.CONTOURS_MATCH_I1,0)
```

```
d2 = cv2.matchShapes(im1,im2,cv2.CONTOURS_MATCH_I2,0)
d3 = cv2.matchShapes(im1,im2,cv2.CONTOURS_MATCH_I3,0)
```

如果上述距离很小，则两个图像（im1 和 im2）是相似的。可以使用任何距离度量，它们通常会产生类似的结果。

1.25 找到 blob 的中心

圆形、正方形、三角形、椭圆形等标准形状的中心相对容易找到。但是当要找到任意形状的质心时，方法并不简单。本节将首先讨论如何找到任意形状的 blob 的中心，然后讨论多个 blob 的情况。

blob 是图像中一组连接的像素，它们具有一些共同的属性（如灰度值）。本节的目标是在 Python 中使用 OpenCV 找到二进制 blob 的中心。如果感兴趣的形状不是二进制的，则必须先对其进行二进制化。

形状的质心是形状中所有点的算术平均值。假设一个形状由 n 个不同的点 $x_1, x_2, ..., x_n$ 组成，则质心由下式给出：

$$c = \frac{1}{n}\sum_{i=1}^{n} x_i$$

在图像处理和计算机视觉的背景下，每个形状都是由像素组成的，质心只是构成形状的所有像素的加权平均值。

可以使用 OpenCV 中的矩找到 blob 的中心。但首先，应该知道图像矩到底是什么。图像矩是图像像素强度的特定加权平均值，借助它可以找到图像的一些特定属性，如半径、面积、质心等。为了找到图像的质心，通常将其转换为二进制格式，然后找到它的中心。

质心由下列公式给出：

$$c_x = \frac{M_{10}}{M_{00}}$$

$$c_y = \frac{M_{01}}{M_{00}}$$

式中，c_x 是 x 坐标；c_y 是质心的 y 坐标；M 表示矩。

执行以下步骤找到 blob 的中心：
（1）将图像转换为灰度。
（2）对图像进行二值化。
（3）计算矩后找到图像的中心。

找图像中单个 blob 的中心的代码如下：

```
# convert image to grayscale image
gray_image = cv2.cvtColor(img, cv2.COLOR_BGR2GRAY)

# convert the grayscale image to binary image
ret,thresh = cv2.threshold(gray_image,127,255,0)
```

```python
# calculate moments of binary image
M = cv2.moments(thresh)

# calculate x,y coordinate of center
cX = int(M["m10"] / M["m00"])
cY = int(M["m01"] / M["m00"])

# put text and highlight the center
cv2.circle(img, (cX, cY), 5, (255, 255, 255), -1)
cv2.putText(img, "centroid", (cX - 25, cY - 25),cv2.FONT_HERSHEY_SIMPLEX, 0.5, (255, 255, 255), 2)

# display the image
cv2.imshow("Image", img)
cv2.waitKey(0)
```

只找到一个 blob 的中心很容易,但是如果图像中有多个 blob 怎么办?那么,将不得不使用 findContours()函数来查找图像中轮廓的数量并找到每个轮廓的中心。下面通过代码看看它是如何工作的:

```python
# read image through command line
img = cv2.imread(args["ipimage"])

# convert the image to grayscale
gray_image = cv2.cvtColor(img, cv2.COLOR_BGR2GRAY)

# convert the grayscale image to binary image
ret,thresh = cv2.threshold(gray_image,127,255,0)

# find contours in the binary image
im2, contours, hierarchy = cv2.findContours(thresh,cv2.RETR_TREE,cv2.CHAIN_APPROX_SIMPLE)
for c in contours:
    # calculate moments for each contour
    M = cv2.moments(c)

    # calculate x,y coordinate of center
    cX = int(M["m10"] / M["m00"])
    cY = int(M["m01"] / M["m00"])
    cv2.circle(img, (cX, cY), 5, (255, 255, 255), -1)
    cv2.putText(img, "centroid", (cX - 25, cY - 25),cv2.FONT_HERSHEY_SIMPLEX, 0.5, (255, 255, 255), 2)

    # display the image
    cv2.imshow("Image", img)
    cv2.waitKey(0)
```

1.26 查找凸包

本节讲解如何找到一个形状（一组点）的凸包。

什么是凸包？让我们将这个术语分解为两部分——凸面和外壳。凸物体是内角不大于180°的物体，而非凸物体称为非凸形或凹形。凸面指一个多边型，没有凹的地方。外壳是指物体的外部或形状。

因此，一个形状或一组点的凸包是围绕点或形状的紧密拟合凸边界。

如果给定一组定义形状的点，那么，如何找到它的凸包？寻找凸包的算法通常被称为礼品包装算法。

如何在图像上使用礼品包装算法？首先需要对正在使用的图像进行二值化，找到轮廓，最后找到凸包。让我们一步一步来。

首先读取输入图像。

```
src = cv2.imread("sample.jpg", 1) # read input image
```

然后对输入的图像进行二值化。

分三步执行二值化：

（1）将图像转换为灰度图。

（2）对灰度图进行模糊处理以消除噪声。

（3）用图像阈值法实现二值化。

代码如下。

```
gray = cv2.cvtColor(src, cv2.COLOR_BGR2GRAY) # convert to grayscale
blur = cv2.blur(gray, (3, 3)) # blur the image
ret, thresh = cv2.threshold(blur, 50, 255, cv2.THRESH_BINARY)
```

上面已将图像转换为二进制大对象。接下来需要找到这些二进制大对象的边界。

接下来，使用OpenCV中的findContour()函数找到轮廓。找到的轮廓提供了每个二进制大对象周围的边界点列表。

初学者可能会想：为什么不简单地使用边缘检测？边缘检测会简单地给出边缘的位置。但理想的是找出边是如何相互连接的。函数findContour()可以找到这些连接并将形成轮廓的点作为列表返回。

```
# Finding contours for the thresholded image
im2, contours, hierarchy = cv2.findContours(thresh, cv2.RETR_TREE, cv2.CHAIN_APPROX_SIMPLE)
```

由于我们已经找到了轮廓，现在可以找到每个轮廓的凸包。这可以使用convexHull()函数来完成。

```
# create hull array for convex hull points
hull = []

# calculate points for each contour
for i in range(len(contours)):
    # creating convex hull object for each contour
    hull.append(cv2.convexHull(contours[i], False))
```

最后一步是可视化刚刚找到的凸包。当然，凸包本身也只是一个轮廓，因此可以使用

OpenCV 库中的 drawContours()函数。

```
# create an empty black image
drawing = np.zeros((thresh.shape[0], thresh.shape[1], 3), np.uint8)

# draw contours and hull points
for i in range(len(contours)):
    color_contours = (0, 255, 0) # green - color for contours
    color = (255, 0, 0) # blue - color for convex hull
    # draw ith contour
    cv2.drawContours(drawing, contours, i, color_contours, 1, 8, hierarchy)
    # draw ith convex hull object
    cv2.drawContours(drawing, hull, i, color, 1, 8)
```

1.27 将一个三角形扭曲为另一个三角形

本节讲解如何将图像中的单个三角形扭曲为另一个图像中的另一个三角形。

在计算机图形学中，人们一直在处理扭曲的三角形，因为任何 3D 表面都可以用三角形来近似。图像可以分解成三角形并变形。但是，在 OpenCV 中，没有现成的方法可以将一个三角形内的像素扭曲为另一个三角形内的像素。

现在已知道，要将一个三角形扭曲为另一个三角形，需要使用仿射变换。在 OpenCV 中，warpAffine()函数允许对图像应用仿射变换，但不能对图像内部的三角形区域应用仿射变换。为了打破这个限制，可以在源三角形周围找到一个边界框，并从源图像中裁剪矩形区域。然后对裁剪后的图像应用仿射变换以获得输出图像。上一步至关重要，因为它允许将仿射变换应用于图像的一个小区域，从而提高计算性能。最后，通过用白色填充输出三角形内的像素来创建一个三角形蒙版。当与输出图像相乘时，此蒙版会将三角形外的所有像素变为黑色，同时保留三角形内所有像素的颜色。

在进入细节之前，先读入输入和输出图像，并定义输入和输出三角形。之前的输出图像只是白色的，但可以通过定义以另一幅图像进行读取。

```
# Read input image and convert to float
img1 = cv2.imread("robot.jpg")

# Output image is set to white
img2 = 255 * np.ones(img_in.shape, dtype = img_in.dtype)

# Define input and output triangles
tri1 = np.float32([[[360,200], [60,250], [450,400]]])
tri2 = np.float32([[[400,200], [160,270], [400,400]]])
```

输入和输出现在已经定义好了，并已经准备好完成将输入三角形内的所有像素转换为输出三角形所需的步骤。

1. 计算边界框

在这一步计算三角形周围的边界框。这个想法是为了提高效率只扭曲图像的一小部分而不是整幅图像。

```
# Find bounding box.
r1 = cv2.boundingRect(tri1)
r2 = cv2.boundingRect(tri2)
```

2. 裁剪图像和更改坐标

为了有效地将仿射变换应用于图像的一部分而不是整幅图像，可以根据上一步计算的边界框裁剪输入图像。三角形的坐标也需要修改，以反映它们在新裁剪图像中的位置。这是通过从三角形的 *x* 和 *y* 坐标中减去边界框左上角的 *x* 和 *y* 坐标来完成的。

```
# Offset points by left top corner of the respective rectangles
tri1Cropped = []
tri2Cropped = []

for i in xrange(0, 3):
    tri1Cropped.append(((tri1[0][i][0] - r1[0]),(tri1[0][i][1] - r1[1])))
    tri2Cropped.append(((tri2[0][i][0] - r2[0]),(tri2[0][i][1] - r2[1])))

# Crop input image
img1Cropped = img1[r1[1]:r1[1] + r1[3], r1[0]:r1[0] + r1[2]]
```

3. 估计仿射变换

上面刚刚获得了裁剪后的输入和输出图像中输入和输出三角形的坐标。使用这两个三角形，可以找到仿射变换。该变换将使用以下代码将输入三角形转换为裁剪图像中的输出三角形。

```
# Given a pair of triangles, find the affine transform
warpMat = cv2.getAffineTransform( np.float32(tri1Cropped), np.float32(tri2Cropped) )
```

4. 边界框内扭曲像素

将上一步找到的仿射变换应用于裁剪后的输入图像，以获得裁剪后的输出图像。在 OpenCV 中，可以使用 warpAffine() 函数对图像应用仿射变换。

```
# Apply the Affine Transform just found to the src image
img2Cropped = cv2.warpAffine( img1Cropped, warpMat, (r2[2], r2[3]), None,
flags=cv2.INTER_LINEAR, borderMode=cv2.BORDER_REFLECT_101 )
```

5. 屏蔽三角形外的像素

在上一步已获得了输出矩形图像，但是还需要矩形区域内的三角形。因此，使用 fillConvexPoly() 函数创建了一个遮罩，用于将三角形外的所有像素涂黑。

```
# Get mask by filling triangle
mask = np.zeros((r2[3], r2[2], 3), dtype = np.float32)
cv2.fillConvexPoly(mask, np.int32(tri2Cropped), (1.0, 1.0, 1.0), 16, 0);

img2Cropped = img2Cropped * mask

# Copy triangular region of the rectangular patch to the output image
img2[r2[1]:r2[1]+r2[3], r2[0]:r2[0]+r2[2]] = img2[r2[1]:r2[1]+r2[3], r2[0]:r2[0]+r2[2]]
* ( (1.0, 1.0, 1.0) - mask )

img2[r2[1]:r2[1]+r2[3], r2[0]:r2[0]+r2[2]] = img2[r2[1]:r2[1]+r2[3], r2[0]:r2[0]+r2[2]]
+ img2Cropped
```

1.28 阿尔法混合

本节讲解如何在 OpenCV 中使用阿尔法混合将透明的 PNG 格式图像覆盖在另一幅图像上。

阿尔法混合是将具有透明度的前景图像覆盖在背景图像上的过程。透明度通常是图像（例如在透明的 PNG 格式的图像中）的第 4 个通道，但它也可以是单独的图像。这种透明蒙版通常称为阿尔法蒙版或阿尔法遮罩。

阿尔法混合的数学表达是直截了当的：

$$I=\alpha F+(1-\alpha)B \quad 0 \leqslant \alpha \leqslant 1$$

式中，α 为阿尔法掩码；F 为前景图像的颜色；B 为背景图像的颜色。

该公式表明，在图像的每像素处，都需要使用阿尔法掩码（α）组合前景图像的颜色（F）和背景图像的颜色（B）。

注意：公式中使用的 α 值实际上是阿尔法掩码中的像素值除以 255，所以在公式中 $0 \leqslant \alpha \leqslant 1$。根据上面的等式，可以进行以下观察。

（1）当 $\alpha=0$ 时，输出像素颜色为背景。

（2）当 $\alpha=1$ 时，输出像素颜色只是前景。

（3）当 $0<\alpha<1$ 时，输出像素颜色是背景和前景的混合。对于逼真的混合，阿尔法蒙版的边界通常具有介于 0 和 1 之间的像素。

让我们通过一个 Python 的例子，看看如何在 OpenCV 中实现阿尔法混合。

```python
import cv2

# Read the images
foreground = cv2.imread("puppets.png")
background = cv2.imread("ocean.png")
alpha = cv2.imread("puppets_alpha.png")

# Convert uint8 to float
foreground = foreground.astype(float)
background = background.astype(float)

# Normalize the alpha mask to keep intensity between 0 and 1
alpha = alpha.astype(float)/255

# Multiply the foreground with the alpha matte
foreground = cv2.multiply(alpha, foreground)

# Multiply the background with ( 1 - alpha )
background = cv2.multiply(1.0 - alpha, background)

# Add the masked foreground and background
outImage = cv2.add(foreground, background)

# Display image
cv2.imshow("outImg", outImage/255)
cv2.waitKey(0)
```

1.29 基于特征的图像对齐

本节讲解如何使用 OpenCV 实现基于特征的图像对齐。

下面通过一个示例来演示这些步骤。在该示例中，将完成使用手机拍摄的表格照片与表格模板对齐。对此所使用的技术通常被称为"基于特征的"图像对齐。因为在这种技术中，需要在一幅图像中检测到一组稀疏的特征，并与另一幅图像中的特征进行匹配，然后基于将一幅图像扭曲到另一幅图像的这些匹配特征计算转换。

在许多应用程序中，有两个相同场景或相同文档的图像，但它们没有对齐。换句话说，如果在一幅图像上选择一个特征（如一个角），在另一幅图像中同一个角的坐标是非常不同的。

图像对齐（也称为图像配准）是一种扭曲一幅图像（或有时是两幅图像）的技术，以便两幅图像中的特征完美对齐。

图像对齐有许多应用。

在许多文档处理应用程序中，第一步是将扫描或拍照的文档与模板对齐。例如，如果想编写一个自动表单阅读器，最好先将表单与其模板对齐，然后根据模板中的固定位置读取字段。

图像对齐最有趣的应用可能是创建全景图。在这种情况下，这两幅图像不是平面图像，而是 3D 场景图像。通常，3D 对齐需要深度信息。但是，当通过围绕其光轴旋转相机拍摄两幅图像时（如全景图的情况），可以使用本节中描述的技术来对齐全景图的两幅图像。

基于特征的图像对齐步骤如下：

（1）读取图像：读取参考图像（或模板图像）和想要与此模板对齐的图像。

（2）检测特征：检测两幅图像中的 ORB 特征。虽然只需要 4 个特征来计算单应性，但通常在两幅图像中检测到数百个特征。可以使用参数 MAX_FEATURES 来控制特征的数量。

（3）匹配特征：在两幅图像中找到匹配的特征，按照匹配度对它们进行排序，并且只保留一小部分原始匹配。

（4）计算单应性：当在两幅图像中有 4 个或更多对应点时，可以计算单应性。自动特征匹配并不总是产生 100%准确的匹配，而 20%～30%的匹配错误并不罕见。幸运的是，findHomography()方法利用了一种称为随机样本共识（RANSAC）的稳健估计技术，即使在存在大量错误匹配的情况下也能产生正确的结果。

（5）扭曲图像：一旦计算出准确的单应性，就可以将变换应用于一幅图像中的所有像素，以将其映射到另一幅图像。这是使用 OpenCV 中的 warpPerspective()函数完成的。

图像配准的 Python 代码如下：

```python
from __future__ import print_function
import cv2
import numpy as np

MAX_FEATURES = 500
GOOD_MATCH_PERCENT = 0.15


def alignImages(im1, im2):

  # Convert images to grayscale
  im1Gray = cv2.cvtColor(im1, cv2.COLOR_BGR2GRAY)
  im2Gray = cv2.cvtColor(im2, cv2.COLOR_BGR2GRAY)
```

```python
# Detect ORB features and compute descriptors
orb = cv2.ORB_create(MAX_FEATURES)
keypoints1, descriptors1 = orb.detectAndCompute(im1Gray, None)
keypoints2, descriptors2 = orb.detectAndCompute(im2Gray, None)

# Match features
matcher = cv2.DescriptorMatcher_create(cv2.DESCRIPTOR_MATCHER_BRUTEFORCE_HAMMING)
matches = matcher.match(descriptors1, descriptors2, None)

# Sort matches by score
matches.sort(key=lambda x: x.distance, reverse=False)

# Remove not so good matches
numGoodMatches = int(len(matches) * GOOD_MATCH_PERCENT)
matches = matches[:numGoodMatches]

# Draw top matches
imMatches = cv2.drawMatches(im1, keypoints1, im2, keypoints2, matches, None)
cv2.imwrite("matches.jpg", imMatches)

# Extract location of good matches
points1 = np.zeros((len(matches), 2), dtype=np.float32)
points2 = np.zeros((len(matches), 2), dtype=np.float32)

for i, match in enumerate(matches):
    points1[i, :] = keypoints1[match.queryIdx].pt
    points2[i, :] = keypoints2[match.trainIdx].pt

# Find homography
h, mask = cv2.findHomography(points1, points2, cv2.RANSAC)

# Use homography
height, width, channels = im2.shape
im1Reg = cv2.warpPerspective(im1, h, (width, height))

return im1Reg, h

if __name__ == '__main__':

    # Read reference image
    refFilename = "form.jpg"
    print("Reading reference image : ", refFilename)
    imReference = cv2.imread(refFilename, cv2.IMREAD_COLOR)

    # Read image to be aligned
    imFilename = "scanned-form.jpg"
    print("Reading image to align : ", imFilename);
    im = cv2.imread(imFilename, cv2.IMREAD_COLOR)
```

```
print("Aligning images ...")
# Registered image will be resotred in imReg
# The estimated homography will be stored in h
imReg, h = alignImages(im, imReference)

# Write aligned image to disk
outFilename = "aligned.jpg"
print("Saving aligned image : ", outFilename);
cv2.imwrite(outFilename, imReg)

# Print estimated homography
print("Estimated homography : \n", h)
```

1.30 使用 ZBar 和 OpenCV 编写条形码和二维码扫描仪的 Python 代码

本节介绍使用名为 ZBar 的库和 OpenCV 编写条形码和二维码扫描仪的 Python 代码。

因为 ZBar 的正式版本不支持 Python 3，所以推荐使用同时支持 Python 2 和 Python 3 的 pyzbar。pyzbar 通过 Python 2 和 Python 3 接口，使用 Zbar 库读取条形码。

安装 pyzbar：

```
pip install pyzbar
```

ZBar 返回的一个条形码/二维码对象有 3 个字段：

（1）类型：如果 ZBar 检测到的符号是二维码，则类型为二维码。如果是条形码，则类型是 ZBar 能够读取的几种条码中的一种。

（2）数据：这是嵌入在条形码/二维码中的数据。此数据通常是字母数字，但其他类型（数字、字节/二进制数字等）也是有效的。

（3）位置：这是定位代码的点的集合。对于 QR（Quick Response）二维码，它是对应于 QR 二维码四边形的 4 个角的 4 个点的列表。对于条形码，位置是标记条形码中单词边界开始和结束的点的集合。

pyzbar 有一个简单的解码函数用来定位和解码图像中的所有符号。解码函数简单地包装了 pyzbar 的 decode() 函数并循环定位到的条形码和 QR 二维码并打印数据。

上一步的解码符号被传递给显示函数。如果这些点形成一个四边形（例如在二维码中），则只需绘制四边形。如果位置不是四边形，就绘制所有点的外边界（也称为凸包）。这是使用 OpenCV 中名为 cv2.convexHull() 的函数完成的。

```
from __future__ import print_function
import pyzbar.pyzbar as pyzbar
import numpy as np
import cv2

def decode(im) :
  # Find barcodes and QR codes
  decodedObjects = pyzbar.decode(im)
```

```python
  # Print results
  for obj in decodedObjects:
    print('Type : ', obj.type)
    print('Data : ', obj.data,'\n')

  return decodedObjects

# Display barcode and QR code location
def display(im, decodedObjects):

  # Loop over all decoded objects
  for decodedObject in decodedObjects:
    points = decodedObject.polygon

    # If the points do not form a quad, find convex hull
    if len(points) > 4 :
      hull = cv2.convexHull(np.array([point for point in points], dtype=np.float32))
      hull = list(map(tuple, np.squeeze(hull)))
    else :
      hull = points;

    # Number of points in the convex hull
    n = len(hull)

    # Draw the convext hull
    for j in range(0,n):
      cv2.line(im, hull[j], hull[ (j+1) % n], (255,0,0), 3)

  # Display results
  cv2.imshow("Results", im);
  cv2.waitKey(0);

# Main
if __name__ == '__main__':

  # Read image
  im = cv2.imread('zbar-test.jpg')

  decodedObjects = decode(im)
  display(im, decodedObjects)
```

1.31 换脸

本节通过 8 个简单的步骤来解释如何使用 OpenCV 和 Python 进行人脸交换。
1）拍摄两幅图像
"源图像"是从所拍摄的两幅图像中提取的人脸图像；"目标图像"是放置从源图像中提取

的人脸图像的地方。

```
img = cv2.imread("bradley_cooper.jpg")
img_gray = cv2.cvtColor(img, cv2.COLOR_BGR2GRAY)
img2 = cv2.imread("jim_carrey.jpg")
img2_gray = cv2.cvtColor(img2, cv2.COLOR_BGR2GRAY)
```

2)找到两幅图像的地标点

使用 dlib 库来检测面部的地标点。

在下面的代码中,展示了如何找到地标点。

在展示的这个特定代码中,检测了源图像的地标,但还需要将其应用于目标图像。

```
# We load Face detector and Face landmarks predictor
detector = dlib.get_frontal_face_detector()
predictor = dlib.shape_predictor("shape_predictor_68_face_landmarks.dat")

# Face 1
faces = detector(img_gray)
for face in faces:
    landmarks = predictor(img_gray, face)
    landmarks_points = []
    for n in range(0, 68):
        x = landmarks.part(n).x
        y = landmarks.part(n).y
        landmarks_points.append((x, y))
```

3)三角形剖分源图像

三角形剖分源图像即把脸分割成三角形。这一步是人脸交换的核心,后面将简单地将每个三角形与目标图像对应的三角形交换。

为什么要把脸分成三角形?

不能只是从源图像中剪下脸部并将其放入目标图像中,因为它们具有不同的大小和视角;也不能马上改变它的大小和视角,因为脸会失去原来的比例。相反,如果将脸部分割成三角形,则可以简单地交换每个三角形,这样它会保持比例,并且会匹配新脸部的表情,如微笑、闭上眼睛或张开嘴。

在下面的代码中,将看到如何使用 OpenCV 进行 Delaunay 算法三角形剖分。

```
# Delaunay triangulation
rect = cv2.boundingRect(convexhull)
subdiv = cv2.Subdiv2D(rect)
subdiv.insert(landmarks_points)
triangles = subdiv.getTriangleList()
triangles = np.array(triangles, dtype=np.int32)
```

4)三角形剖分目标图像

目标图像的三角形剖分需要与源图像的三角形剖分具有相同的模式。这意味着点的连接必须相同。因此,在对源图像进行三角形剖分之后,可从该三角形剖分中获取地标点的索引,以便在目标图像上复制相同的三角形剖分。

```
# we get the Landmark points indexes of each triangle
indexes_triangles = []
for t in triangles:
```

```
    pt1 = (t[0], t[1])
    pt2 = (t[2], t[3])
    pt3 = (t[4], t[5])

    index_pt1 = np.where((points == pt1).all(axis=1))
    index_pt1 = extract_index_nparray(index_pt1)
    index_pt2 = np.where((points == pt2).all(axis=1))
    index_pt2 = extract_index_nparray(index_pt2)
    index_pt3 = np.where((points == pt3).all(axis=1))
    index_pt3 = extract_index_nparray(index_pt3)

    if index_pt1 is not None and index_pt2 is not None and index_pt3 is not None:
        triangle = [index_pt1, index_pt2, index_pt3]
        indexes_triangles.append(triangle)
```

一旦有了三角形索引，就可以遍历它们并对目标脸进行三角形剖分。

```
# Triangulation destinazion face
for triangle_index in indexes_triangles:
    # Triangulation of the Second face
    tr1_pt1 = landmarks_points2[triangle_index[0]]
    tr1_pt2 = landmarks_points2[triangle_index[1]]
    tr1_pt3 = landmarks_points2[triangle_index[2]]
    triangle2 = np.array([tr1_pt1, tr1_pt2, tr1_pt3], np.int32)
```

5）提取和扭曲三角形

一旦对两幅图像中的人脸进行了三角形剖分，就会获取源人脸的三角形并提取它们，但还需要获取目标人脸三角形的坐标，这样才能将源人脸的三角形扭曲成与目标人脸匹配的三角形。

下面的代码显示了如何扭曲源图像中人脸的三角形。

```
# Warp triangles
points = np.float32(points)
points2 = np.float32(points2)
M = cv2.getAffineTransform(points, points2)
warped_triangle = cv2.warpAffine(cropped_triangle, M, (w, h))
warped_triangle = cv2.bitwise_and(warped_triangle, warped_triangle, mask=cropped_tr2_mask)
```

6）将扭曲的三角形连接在一起

切割并扭曲了所有三角形之后，还需要将它们连接在一起。这是只是使用三角形剖分模式重建面部，唯一的区别是这次放置了扭曲的三角形。在此操作结束时，已准备好更换脸部。

```
# Reconstructing destination face
img2_new_face = np.zeros((1155, 849, 3), np.uint8)
img2_new_face_rect_area = img2_new_face[y: y + h, x: x + w]
img2_new_face_rect_area_gray = cv2.cvtColor(img2_new_face_rect_area, cv2.COLOR_BGR2GRAY)

# Let's create a mask to remove the lines between the triangles
_, mask_triangles_designed = cv2.threshold(img2_new_face_rect_area_gray, 1, 255, cv2.THRESH_BINARY_INV)
warped_triangle = cv2.bitwise_and(warped_triangle, warped_triangle, mask=mask_
```

```
triangles_designed)

    img2_new_face_rect_area = cv2.add(img2_new_face_rect_area, warped_triangle)
    img2_new_face[y: y + h, x: x + w] = img2_new_face_rect_area
```

7) 替换目标图像上的人脸

现在可以更换脸部了。先剪掉目标图像中的脸,为新脸腾出空间。所以,需要把新面孔和没有面孔的目标图像连接在一起。

```
# Face swapped (putting 1st face into 2nd face)
img2_face_mask = np.zeros_like(img2_gray)
img2_head_mask = cv2.fillConvexPoly(img2_face_mask, convexhull2, 255)
img2_face_mask = cv2.bitwise_not(img2_head_mask)

img2_head_noface = cv2.bitwise_and(img2, img2, mask=img2_face_mask)
result = cv2.add(img2_head_noface, img2_new_face)
```

8) 无缝克隆

至此,面部被正确交换,是时候调整颜色以使源图像适合目标图像了。在 OpenCV 中,有一个名为 seamlessClone() 的内置函数,它会自动执行此操作。

我们需要获取新的脸(已在步骤 6)创建,获取原始的目标图像和它的蒙版来切出脸,需要获得脸的中心。

```
(x, y, w, h) = cv2.boundingRect(convexhull2)
center_face2 = (int((x + x + w) / 2), int((y + y + h) / 2))

seamlessclone = cv2.seamlessClone(result, img2, img2_head_mask, center_face2, cv2.MIXED_CLONE)
```

1.32 applyColorMap 用于伪着色

假设想在地图上显示美国不同地区的温度,可以将温度数据叠加在美国地图上作为灰度图像——较暗的区域代表较冷的温度,较亮的区域代表较热的区域。这种表示由于两个主要原因,它是一种糟糕的表示。首先,人类视觉系统没有优化到测量灰度强度的微小变化,而是更善于感知颜色的变化;其次,将不同的含义与不同的颜色联系起来。用蓝色表示较冷的温度,用红色表示较暖的温度,这更有意义。

温度数据只是一个示例,但还有其他几种数据是单值(灰度)的情况,但将其转换为颜色数据以进行可视化是有意义的。其他可以通过伪着色更好地可视化的数据是高度、压力、密度、湿度等。

OpenCV 定义了 12 种颜色图,可以使用函数 applyColorMap() 将其应用于灰度图像以生成伪彩色图像。下面的代码可以说明如何将颜色图 COLORMAP_JET 应用于图像。

```
import cv2

im_gray = cv2.imread("pluto.jpg", cv2.IMREAD_GRAYSCALE)
im_color = cv2.applyColorMap(im_gray, cv2.COLORMAP_JET)
```

1.33 高动态范围成像

本节讲解如何使用不同曝光设置拍摄的多幅图像来创建高动态范围（HDR）图像。

大多数数码相机和显示器将彩色图像捕获或显示为 24 位矩阵。每个颜色通道有 8 位，因此每个通道的像素值为 0～255。换句话说，普通相机或显示器具有有限的动态范围。

然而，人们周围的世界有着非常大的动态范围。例如，车库里关灯时会变得漆黑一片，如果人直视太阳，就会觉得非常亮。即使不考虑这些极端情况，在日常情况下，8 位也勉强可以捕捉场景。因此，相机会尝试估计光照并自动设置曝光，以使图像中最有趣的部分具有良好的动态范围，并且将太暗和太亮的部分分别裁剪为 0 和 255。

高动态范围成像如何工作？

当使用相机拍照时，相机的每个颜色通道只有 8 位来表示场景的动态范围（亮度范围）。但是可以通过改变快门速度在不同曝光下拍摄多张相同场景的照片。大多数单反相机都有一个被称为自动包围曝光（AEB）的功能，只需按一下按钮，就可以在不同的曝光下拍摄多张照片。

使用相机上的 AEB 或手机上的自动包围式应用程序，都可以一张一张地快速拍摄多张照片，这样场景就不会改变。当在 iPhone 中使用 HDR 成像模式时，iPhone 会拍摄 3 张照片。

（1）曝光不足的图像：此图像比正确曝光的图像更暗。目标是捕获图像中非常明亮的部分。

（2）正确曝光的图像：这是相机根据其估计的照度拍摄的常规图像。

（3）曝光过度的图像：此图像比正确曝光的图像更亮。目标是捕捉图像中非常暗的部分。

但是，如果场景的动态范围非常大，则可以拍摄 3 张以上的照片来合成 HDR 图像。本节将使用 4 张曝光时间分别为 1/30、0.25、2.5 和 15 秒的照片。

单反相机或手机使用的有关曝光时间和其他设置的信息通常存储在 JPEG 文件的 EXIF 元数据中。可以使用 EXIF 命令行实用程序 EXIFTOOL 查看存储在 Windows 或 macos 中的 JPEG 文件中的 EXIF 元数据。

先从读取分配不同曝光时间的照片开始。

```
def readImagesAndTimes():
  # List of exposure times
  times = np.array([ 1/30.0, 0.25, 2.5, 15.0 ], dtype=np.float32)

  # List of image filenames
  filenames = ["img_0.033.jpg", "img_0.25.jpg", "img_2.5.jpg", "img_15.jpg"]
  images = []
  for filename in filenames:
    im = cv2.imread(filename)
    images.append(im)

  return images, times
```

用于组成 HDR 图像的照片未对齐会导致严重的伪影。

当然，在拍摄用于创建 HDR 图像的照片时，专业摄影师会将相机安装在三脚架上。他们还使用称为反光镜锁定的功能来减少额外的振动。即使这样，照片也可能无法完美对齐，因为无法保证无振动的环境。当使用手持相机或手机拍摄照片时，对齐问题会变得更糟。

幸运的是，OpenCV 提供了一种使用 AlignMTB 算法对齐这些照片的简单方法。该算法将所有照片转换为中值阈值位图（MTB）。照片的 MTB 是通过将值 1 分配给比中值亮度更亮的像素，否则用值为 0 来计算的。MTB 对曝光时间是不变的。因此，MTB 可以对齐，而无须指定曝光时间。

使用以下代码行执行基于 MTB 的对齐。

```
# Align input images
alignMTB = cv2.createAlignMTB()
alignMTB.process(images, images)
```

典型相机的响应对场景亮度不是线性的。这意味着什么？假设，两个物体被相机拍摄，其中一个在现实世界中的亮度是另一个的 2 倍。当测量照片中 2 个物体的像素强度时，较亮物体的像素值不会是较暗物体的 2 倍！如果不估计相机响应函数（CRF），则将无法将照片合并为一幅 HDR 图像。

将多张曝光照片合并为 HDR 图像意味着什么？

只考虑图像某个位置 (x, y) 的一像素。如果 CRF 是线性的，则像素值将与曝光时间成正比，除非像素在特定图像中太暗（即接近 0）或太亮（即接近 255）。可以过滤掉这些坏像素（太暗或太亮），并通过将像素值除以曝光时间来估计像素处的亮度，然后在像素不错的所有图像中平均该亮度值。可以对所有像素执行此操作并获得单个图像，其中所有像素都是通过平均"好"像素获得的。

但是 CRF 不是线性的，需要先使图像强度线性化，然后才能通过首先估计 CRF 来合并/平均它们。

好消息是，如果知道每张照片的曝光时间，就可以从中估计 CRF。与计算机视觉中的许多问题一样，寻找 CRF 的问题被设置为一个优化问题，其目标是最小化由数据项和平滑项组成的目标函数。这些问题通常简化为线性最小二乘问题，使用奇异值分解（SVD）来解决。SVD 是所有线性代数包的一部分。

在 OpenCV 中使用 CalibrateDebevec 或 CalibrateRobertson 只需 2 行代码即可找到 CRF。本节使用 CalibrateDebevec。

```
# Obtain Camera Response Function (CRF)
CalibrateDebevec = cv2.createCalibrateDebevec()
responseDebevec = calibrateDebevec.process(images, times)
```

估计 CRF 后，可以使用 MergeDebevec 将曝光照片合并为一幅 HDR 图像。

```
# Merge images into an HDR linear image
mergeDebevec = cv2.createMergeDebevec()
hdrDebevec = mergeDebevec.process(images, times, responseDebevec)
# Save HDR image
cv2.imwrite("hdrDebevec.hdr", hdrDebevec)
```

上面保存的 HDR 图像可以在 Photoshop 中加载并进行色调映射。

现在已将曝光照片合并为一幅 HDR 图像。你能猜出这幅图像的最小和最大像素值吗？对于漆黑的情况，最小值显然是 0。理论最大值是多少？无限！在实践中，不同情况下的最大值是不同的。如果场景包含非常亮的光源，将看到一个非常大的最大值。

尽管前面已经使用多张照片恢复了相对的亮度信息，但现在面临着将这些信息保存为 24 位图像以供显示的挑战。

将 HDR 图像转换为每通道 8 位图像同时保留尽可能多的细节的过程称为色调映射。

在 OpenCV 中可实现 Drago、Durand、Reinhard 和 Mantiuk 4 个色调映射算法。

（1）不同色调映射算法的常用参数。

①gamma：通过应用 gamma 校正来压缩动态范围。当 gamma=1 时，不应用校正；当 gamma<1 时图像会变暗；当 gamma>1 时图像会变亮。

②saturation：用于增加或减少饱和度。饱和度高时，颜色更丰富、更强烈；饱和度值更接近于零，则颜色逐渐消失为灰度。

③contrast：控制输出图像的对比度。

（2）OpenCV 中可用的 4 个色调映射算法。

①Drago 色调映射算法。

Drago 色调映射算法的参数如下：

```
createTonemapDrago
(
  float   gamma = 1.0f,
  float   saturation = 1.0f,
  float   bias = 0.85f
)
```

这里，bias 是偏置函数在[0, 1]区间的值。0.7~0.9 的值通常给出最好的结果。默认值为 0.85。

Drago 色调映射算法的 Python 代码如下：

```python
# Tonemap using Drago's method to obtain 24-bit color image
tonemapDrago = cv2.createTonemapDrago(1.0, 0.7)
ldrDrago = tonemapDrago.process(hdrDebevec)
ldrDrago = 3 * ldrDrago
cv2.imwrite("ldr-Drago.jpg", ldrDrago * 255)
```

②Durand 色调映射算法。

Durand 色调映射算法的参数如下：

```
createTonemapDurand
(
  float   gamma = 1.0f,
  float   contrast = 4.0f,
  float   saturation = 1.0f,
  float   sigma_space = 2.0f,
  float   sigma_color = 2.0f
);
```

该算法将图像分解为基础层和细节层。基础层是使用被称为双边滤波器的边缘保持滤波器获得的。sigma_space 和 sigma_color 是双边滤波器的参数，分别控制空间和颜色域中的平滑量。

Durand 色调映射算法的 Python 代码如下：

```python
# Tonemap using Durand's method obtain 24-bit color image
tonemapDurand = cv2.createTonemapDurand(1.5,4,1.0,1,1)
ldrDurand = tonemapDurand.process(hdrDebevec)
ldrDurand = 3 * ldrDurand
cv2.imwrite("ldr-Durand.jpg", ldrDurand * 255)
```

③Reinhard 色调映射算法。

Reinhard 色调映射算法的参数如下：

```
createTonemapReinhard
(
  float   gamma = 1.0f,
  float   intensity = 0.0f,
  float   light_adapt = 1.0f,
  float   color_adapt = 0.0f
)
```

参数 intensity 应在[-8, 8]区间。更大的 intensity 值会产生更亮的结果。参数 light_adapt 控制光照适应，在[0, 1]区间。值为 1 表示仅基于像素值的自适应；值为 0 表示全局自适应。中间值可用于两者的加权组合。参数 color_adapt 控制色度适应，在[0, 1]区间。如果该值设置为 1，则通道被独立处理；如果该值设置为 0，则每个通道的适应级别相同；中间值可用于两者的加权组合。

Reinhard 色调映射算法的 Python 代码如下：

```
# Tonemap using Reinhard's method to obtain 24-bit color image
tonemapReinhard = cv2.createTonemapReinhard(1.5, 0,0,0)
ldrReinhard = tonemapReinhard.process(hdrDebevec)
cv2.imwrite("ldr-Reinhard.jpg", ldrReinhard * 255)
```

④Mantiuk 色调映射算法。

Mantiuk 色调映射算法的参数如下：

```
createTonemapMantiuk
(
  float   gamma = 1.0f,
  float   scale = 0.7f,
  float   saturation = 1.0f
)
```

参数 scale 是对比度的比例因子，其 0.6~0.9 的值产生最佳结果。

Mantiuk 色调映射算法的 Python 代码如下：

```
# Tonemap using Mantiuk's method to obtain 24-bit color image
tonemapMantiuk = cv2.createTonemapMantiuk(2.2,0.85, 1.2)
ldrMantiuk = tonemapMantiuk.process(hdrDebevec)
ldrMantiuk = 3 * ldrMantiuk
cv2.imwrite("ldr-Mantiuk.jpg", ldrMantiuk * 255)
```

1.34 曝光融合

曝光融合是一种将使用不同曝光设置拍摄的照片组合成一个看起来像色调映射的 HDR 图像的方法。

当使用相机拍摄照片时，每个颜色通道只有 8 位来表示场景的亮度。然而，理论上，人们周围世界的亮度可以从 0（漆黑）到几乎无限（直视太阳）不等。因此，傻瓜相机或移动相机根据场景决定曝光设置，以便使用相机的动态范围（0~255 值）来表示照片中最有趣的部分。例如，在许多相机中，人脸检测用于查找人脸，并设置曝光以使人脸看起来光线充足。

这就引出了一个问题——能否在不同的曝光设置下拍摄多张照片并捕捉更大范围的场景亮度？答案是肯定的。传统方法是使用 HDR 图像和色调映射。

HDR 图像需要知道精确的曝光时间。HDR 图像本身看起来很暗，并不漂亮。HDR 图像中的最小强度为 0，但理论上没有最大值。所以，需要将它的值映射到 0~255，这样才能显示它。这种将 HDR 图像映射到每通道 8 位常规彩色图像的过程称为色调映射。

组合 HDR 图像然后进行色调映射有点麻烦。不能只使用多张照片创建一幅色调映射图像而无须使用 HDR。事实证明，可以使用曝光融合做到这一点。

应用曝光融合的步骤如下：

1）捕获具有不同曝光度的多张照片

首先，需要在不移动相机的情况下捕捉同一场景的一系列图像。这是通过改变相机的快门速度来实现的。通常，会选择一些曝光不足的照片和一些曝光过度的照片，以及正确曝光的照片。

在"正确"曝光的照片中，选择快门速度（由相机或摄影者自动选择），以便使用每颜色通道 8 位的动态范围来表示照片中最有趣的部分。太暗的区域被裁剪为 0，太亮的区域饱和到 255。

在曝光不足的照片中，快门速度较快，图像较暗。因此，颜色通道的 8 位用于捕获明亮区域，而黑暗区域被裁剪为 0。

在曝光过度的照片中，快门速度较慢，因此传感器捕获的光线更多，因此图像很亮。传感器颜色通道的 8 位用于捕捉黑暗区域的强度，而明亮区域的饱和度为 255。

大多数单反相机都有一个被称为自动包围曝光（AEB）的功能，只需按一下按钮，就可以在不同的曝光下拍摄多张照片。

一旦捕获了这些图像，就可以使用下面的代码来读取它们。

```python
def readImagesAndTimes():

    filenames = [
                "images/memorial0061.jpg",
                "images/memorial0062.jpg",
                "images/memorial0063.jpg",
                "images/memorial0064.jpg",
                "images/memorial0065.jpg",
                "images/memorial0066.jpg",
                "images/memorial0067.jpg",
                "images/memorial0068.jpg",
                "images/memorial0069.jpg",
                "images/memorial0070.jpg",
                "images/memorial0071.jpg",
                "images/memorial0072.jpg",
                "images/memorial0073.jpg",
                "images/memorial0074.jpg",
                "images/memorial0075.jpg",
                "images/memorial0076.jpg"
                ]

    images = []
    for filename in filenames:
        im = cv2.imread(filename)
```

```
        images.append(im)

    return images
```

2)对齐照片

即使是使用三脚架采集的序列中的照片也需要对齐,因为即使是轻微的相机抖动也会降低最终照片的质量。OpenCV 提供了一种使用 AlignMTB 算法对齐这些照片的简单方法。该算法将所有照片转换为中值阈值位图(MTB)。照片的 MTB 是通过将值为 1 分配给比中值亮度更亮的像素,否则用值为 0 来计算。MTB 对曝光时间是不变的。因此,MTB 可以对齐,而无须指定曝光时间。

使用以下代码执行基于 MTB 的对齐。

```
# Align input images
alignMTB = cv2.createAlignMTB()
alignMTB.process(images, images)
```

不同曝光度的照片捕捉到不同范围的场景亮度。曝光融合通过仅保留多重曝光照片序列中的"最佳"部分来计算所需的图像。这里采用 3 个质量衡量标准:

(1)曝光良好:如果序列照片中的像素接近 0 或接近 255,则不应该使用该照片来找到最终的像素值,而应优选值接近中间强度(128)的像素。

(2)对比度:高对比度通常意味着高质量。因此,对于特定像素的对比度值较高的图像,该像素的权重较高。

(3)饱和度:同样,饱和度越高的颜色越不褪色,代表质量更高的像素。因此,特定像素饱和度高的图像会被赋予该像素更高的权重。

这 3 个质量度量用于创建权重图 $W(i, j, k)$,它表示第 k 幅图像在位置 (i, j) 处像素在最终强度中的贡献。权重图 $W(i, j, k)$ 被归一化,使得对于任何像素 (i, j),所有图像的贡献加起来为 1。

使用 OpenCV,只是使用 MergeMertens 类的两行代码就可以实现曝光融合。

```
mergeMertens = cv2.createMergeMertens()
exposureFusion = mergeMertens.process(images)
```

在输入图像中,在曝光过度的图像中的昏暗区域和曝光不足的图像中的明亮区域获得了细节。然而,在合并的输出图像中,像素都很好地照亮了图像的每个部分的细节。

还可以在上一节中用于 HDR 成像的图像上看到这种效果。

正如读者在本节中所看到的,曝光融合允许在不明确计算 HDR 图像的情况下实现类似于 HDR + Tonemapping 的效果。所以,不需要知道每张照片的曝光时间,就可以获得非常合理的结果。

那么,为什么还需要 HDR 图像的方法呢?因为在许多情况下,曝光融合产生的输出可能不符合一些人的喜好,但又没有旋钮可以调整以使其不同或更好;另一方面,HDR 图像的方法可以捕捉场景的原始亮度。如果不喜欢色调映射的 HDR 图像,还可以尝试不同的色调映射算法。

总之,曝光融合代表了一种权衡——权衡灵活性以支持速度等不太严格的要求,如不需要曝光时间。

1.35　对象跟踪

本节讲解使用 OpenCV 进行对象跟踪。OpenCV 3.0 开始引入跟踪 API。本节如何使用 OpenCV 4.2 中可用的 8 种不同的跟踪器。

简单地说,在视频的连续帧中定位对象称为跟踪。

这个定义听起来很简单,但在计算机视觉和机器学习中,跟踪是一个非常广泛的术语,包含概念上相似但技术上不同的算法。例如,以下所有不同但相关的算法,都是在对象跟踪下所进行的研究。

- 稠密光流:这种算法有助于估计视频帧中每像素的运动矢量。
- 稀疏光流:这种算法,如 KLT (Kanade-Lucas-Tomashi)跟踪算法用于跟踪图像中几个特征点的位置。
- 卡尔曼滤波:一种非常流行的信号处理算法,用于根据先验运动信息预测运动物体的位置。
- Meanshift 和 Camshift:定位密度函数最大值的算法,它们也用于对象跟踪。
- 单个对象跟踪器:在此类跟踪器中,第 1 帧使用矩形标记,以指示要跟踪对象的位置,然后使用跟踪算法在后续帧中跟踪对象。在大多数实际应用中,这些跟踪器与对象检测器结合使用。
- 多对象跟踪查找算法:如果有一个快速对象检测器,那么在每帧中检测多个对象,然后运行跟踪查找算法来识别一帧中的哪个矩形对应于下一帧中的矩形是有意义的。

如果使用 OpenCV 做过人脸检测,那么就会知道它是实时工作的,并且可以轻松地检测每帧中的人脸。那么,为什么首先需要对象跟踪?对象跟踪具有以下优点。

(1) 跟踪比检测快。通常,跟踪算法比检测算法快。原因很简单。当跟踪在前一帧中检测到的对象时,会对对象的外观有很多了解,还知道前一帧中的位置及其运动的方向和速度。所以,可以利用所有这些信息来预测下一帧中物体的位置,并在物体的预期位置周围做一个小搜索,从而准确定位到物体。一个好的跟踪算法将使用它所拥有的关于该对象的所有信息,而检测算法总是从头开始。因此,在设计高效系统时,通常每第 n 帧运行一次对象检测,而在其中间的 $n-1$ 帧中采用跟踪算法。为什么不简单地检测第 1 帧中的对象并随后跟踪它呢?跟踪确实受益于它所拥有的额外信息,但是当物体长时间躲在障碍物后面或者移动速度过快以至于跟踪算法无法赶上时,跟踪算法也可能失去对物体的跟踪。跟踪算法累积误差也很常见,并且跟踪对象的边界框会慢慢偏离它正在跟踪的对象。为了通过跟踪算法解决这些问题,每隔一段时间就会运行一种检测算法。检测算法在大量对象示例上进行训练。因此,它们对对象的一般类别有更多的了解。另一方面,跟踪算法更了解它们正在跟踪的类的具体实例。

(2) 当检测失败时,跟踪可以提供帮助。如果在视频上运行人脸检测器并且人脸被物体遮挡,则人脸检测器很可能会失败。另一方面,一个好的跟踪算法将处理某种程度的遮挡。

(3) 跟踪保留身份信息。对象检测的输出包含对象的矩形数组,但是没有将其标识附加到对象。

OpenCV 4 带有一个跟踪 API,其中包含许多单个对象跟踪算法的实现。OpenCV 4.2 中有 BOOSTING、MIL、KCF、TLD、MEDIANFLOW、GOTURN、MOSSE 和 CSRT 等 8 种不同的跟踪器可用。

在下面的代码中,首先通过选择跟踪器类型来设置跟踪器——BOOSTING、MIL、KCF、

TLD、MEDIAFLOW、GOTURN、MOSSE 或 CSRT。然后打开一个视频并抓取一帧，定义了一个包含第 1 帧对象的边界框，并使用第 1 帧和边界框初始化跟踪器。最后，从视频中读取帧，然后循环更新跟踪器以获得当前帧的新边界框，随后显示结果。

```python
import cv2
import sys

(major_ver, minor_ver, subminor_ver) = (cv2.__version__).split('.')

if __name__ == '__main__' :

    # Set up tracker
    # Instead of MIL, you can also use

    tracker_types = ['BOOSTING', 'MIL','KCF', 'TLD', 'MEDIANFLOW', 'GOTURN', 'MOSSE', 'CSRT']
    tracker_type = tracker_types[2]

    if int(minor_ver) < 3:
        tracker = cv2.Tracker_create(tracker_type)
    else:
        if tracker_type == 'BOOSTING':
            tracker = cv2.TrackerBoosting_create()
        if tracker_type == 'MIL':
            tracker = cv2.TrackerMIL_create()
        if tracker_type == 'KCF':
            tracker = cv2.TrackerKCF_create()
        if tracker_type == 'TLD':
            tracker = cv2.TrackerTLD_create()
        if tracker_type == 'MEDIANFLOW':
            tracker = cv2.TrackerMedianFlow_create()
        if tracker_type == 'GOTURN':
            tracker = cv2.TrackerGOTURN_create()
        if tracker_type == 'MOSSE':
            tracker = cv2.TrackerMOSSE_create()
        if tracker_type == "CSRT":
            tracker = cv2.TrackerCSRT_create()

    # Read video
    video = cv2.VideoCapture("videos/chaplin.mp4")

    # Exit if video not opened
    if not video.isOpened():
        print "Could not open video"
        sys.exit()

    # Read first frame
    ok, frame = video.read()
    if not ok:
        print 'Cannot read video file'
```

```python
        sys.exit()

    # Define an initial bounding box
    bbox = (287, 23, 86, 320)

    # Uncomment the line below to select a different bounding box
    bbox = cv2.selectROI(frame, False)

    # Initialize tracker with first frame and bounding box
    ok = tracker.init(frame, bbox)

    while True:
        # Read a new frame
        ok, frame = video.read()
        if not ok:
            break

        # Start timer
        timer = cv2.getTickCount()

        # Update tracker
        ok, bbox = tracker.update(frame)

        # Calculate Frames per second (FPS)
        fps = cv2.getTickFrequency() / (cv2.getTickCount() - timer);

        # Draw bounding box
        if ok:
            # Tracking success
            p1 = (int(bbox[0]), int(bbox[1]))
            p2 = (int(bbox[0] + bbox[2]), int(bbox[1] + bbox[3]))
            cv2.rectangle(frame, p1, p2, (255,0,0), 2, 1)
        else :
            # Tracking failure
            cv2.putText(frame, "Tracking failure detected", (100,80), cv2.FONT_HERSHEY_SIMPLEX, 0.75,(0,0,255),2)

        # Display tracker type on frame
        cv2.putText(frame, tracker_type + " Tracker", (100,20), cv2.FONT_HERSHEY_SIMPLEX, 0.75, (50,170,50),2);

        # Display FPS on frame
        cv2.putText(frame, "FPS : " + str(int(fps)), (100,50), cv2.FONT_HERSHEY_SIMPLEX, 0.75, (50,170,50), 2);

        # Display result
        cv2.imshow("Tracking", frame)

        # Exit if ESC pressed
        k = cv2.waitKey(1) & 0xff
```

```
            if k == 27 : break
```

1.36 多对象跟踪

本节介绍如何使用通过 MultiTracker 类实现 OpenCV 的多对象跟踪 API。

大多数计算机视觉和机器学习的初学者都学习对象检测。初学者可能会想：为什么需要对象跟踪，不能只检测每帧中的对象吗？

首先探讨一下为什么跟踪很有用的几个原因。

- 当在视频帧中检测到多个对象（例如人）时，跟踪有助于跨帧建立对象的身份。
- 在某些情况下，对象检测可能会失败，但仍然可以跟踪对象，因为跟踪考虑了对象在前一帧中的位置和外观。
- 一些跟踪算法的效率高，因为它们进行局部搜索而不是全局搜索。因此，可以通过对每第 n 帧执行对象检测并在中间帧跟踪对象来为系统获得非常高的帧速率。

那么，为什么不在第一次检测后无限期地跟踪对象呢？跟踪算法有时会丢失它正在跟踪的对象。例如，当对象的运动速度太快时，跟踪算法可能跟不上。许多现实世界的应用程序同时使用检测和跟踪。

本节只关注跟踪部分，要跟踪的对象将通过在它们周围拖动一个边界框来指定。

OpenCV 中的 MultiTracker 类提供了多目标跟踪的实现。这是一个初级的实现，因为它独立地处理被跟踪对象，而不对被跟踪对象进行任何优化。

下面通过一步步地查看代码，了解如何使用 OpenCV 中的多对象跟踪 API。

1）创建单个对象跟踪器

多对象跟踪器只是单个对象跟踪器的集合。首先定义一个函数，该函数将跟踪器类型作为输入并创建一个跟踪器对象。OpenCV 提供了 8 种不同的跟踪器类型：BOOSTING、MIL、KCF、TLD、MEDIAFLOW、GOTURN、MOSSE、CSRT。

在下面的代码中，给定跟踪器类的名称，并返回跟踪器对象。这将在稍后用于填充多跟踪器。

```
from __future__ import print_function
import sys
import cv2
from random import randint

trackerTypes = ['BOOSTING', 'MIL', 'KCF','TLD', 'MEDIANFLOW', 'GOTURN', 'MOSSE', 'CSRT']

def createTrackerByName(trackerType):
  # Create a tracker based on tracker name
  if trackerType == trackerTypes[0]:
    tracker = cv2.TrackerBoosting_create()
  elif trackerType == trackerTypes[1]:
    tracker = cv2.TrackerMIL_create()
  elif trackerType == trackerTypes[2]:
    tracker = cv2.TrackerKCF_create()
  elif trackerType == trackerTypes[3]:
```

```
    tracker = cv2.TrackerTLD_create()
  elif trackerType == trackerTypes[4]:
    tracker = cv2.TrackerMedianFlow_create()
  elif trackerType == trackerTypes[5]:
    tracker = cv2.TrackerGOTURN_create()
  elif trackerType == trackerTypes[6]:
    tracker = cv2.TrackerMOSSE_create()
  elif trackerType == trackerTypes[7]:
    tracker = cv2.TrackerCSRT_create()
  else:
    tracker = None
    print('Incorrect tracker name')
    print('Available trackers are:')
    for t in trackerTypes:
      print(t)

  return tracker
```

2）读取视频的第 1 帧

多目标跟踪器需要两个输入：

（1）一个视频帧。

（2）要跟踪的所有对象的位置（边界框）。

给定这些信息，跟踪器在所有后续帧中跟踪这些指定对象的位置。

在下面的代码中，首先使用了 VideoCapture 类加载视频并读取第 1 帧。这将在稍后用于初始化 MultiTracker。

```
# Set video to load
videoPath = "videos/run.mp4"

# Create a video capture object to read videos
cap = cv2.VideoCapture(videoPath)

# Read first frame
success, frame = cap.read()
# quit if unable to read the video file
if not success:
  print('Failed to read video')
  sys.exit(1)
```

3）在第 1 帧中定位对象

需要在第 1 帧中定位想要跟踪的对象。该位置只是一个边界框。

OpenCV 提供了一个名为 selectROI()的函数，它会弹出一个图形用户界面（GUI）供用户选择边界框。

对于每个对象，还需选择一种随机颜色来显示边界框。

代码如下：

```
## Select boxes
bboxes = []
colors = []
```

```
# OpenCV's selectROI function doesn't work for selecting multiple objects in Python
# So we will call this function in a loop till we are done selecting all objects
while True:
  # draw bounding boxes over objects
  # selectROI's default behaviour is to draw box starting from the center
  # when fromCenter is set to false, you can draw box starting from top left corner
  bbox = cv2.selectROI('MultiTracker', frame)
  bboxes.append(bbox)
  colors.append((randint(0, 255), randint(0, 255), randint(0, 255)))
  print("Press q to quit selecting boxes and start tracking")
  print("Press any other key to select next object")
  k = cv2.waitKey(0) & 0xFF
  if (k == 113):  # q is pressed
    break

print('Selected bounding boxes {}'.format(bboxes))
```

4）初始化 MultiTracker

到目前为止，VideoCapture 已经读取了第 1 帧并获得了对象周围的边界框。这就是初始化多目标跟踪器所需的所有信息。

下面首先创建一个 MultiTracker 对象，并向其中添加与边界框一样多的单个对象跟踪器。在此示例中，使用 CSRT 单对象跟踪器，但也可以通过将下面的 trackerType 变量更改为本文开头提到的 8 个跟踪器之一来尝试其他跟踪器类型。CSRT 跟踪器不是速度最快的，但在许多情况下它产生了最好的结果。

MultiTracker 类只是这些单个对象跟踪器的包装器。单个对象跟踪器是使用第 1 帧初始化的，并且边界框指示了想要跟踪的对象的位置。MultiTracker 将此信息传递给它在内部包装的单个对象跟踪器。

```
# Specify the tracker type
trackerType = "CSRT"

# Create MultiTracker object
multiTracker = cv2.MultiTracker_create()

# Initialize MultiTracker
for bbox in bboxes:
  multiTracker.add(createTrackerByName(trackerType), frame, bbox)
```

5）更新 MultiTracker 并显示结果

MultiTracker 已准备就绪，并且可以在新帧中跟踪多个对象。下面使用 MultiTracker 类的 update()方法来定位新帧中的对象。每个被跟踪对象的每个边界框都使用不同的颜色绘制。代码如下：

```
# Process video and track objects
while cap.isOpened():
  success, frame = cap.read()
  if not success:
    break
```

```
    # get updated location of objects in subsequent frames
    success, boxes = multiTracker.update(frame)

    # draw tracked objects
    for i, newbox in enumerate(boxes):
      p1 = (int(newbox[0]), int(newbox[1]))
      p2 = (int(newbox[0] + newbox[2]), int(newbox[1] + newbox[3]))
      cv2.rectangle(frame, p1, p2, colors[i], 2, 1)

    # show frame
    cv2.imshow('MultiTracker', frame)

    # quit on ESC button
    if cv2.waitKey(1) & 0xFF == 27:  # Esc pressed
   break
```

1.37 自动红眼去除器

本节介绍如何完全自动地从照片中去除红眼。

当人在黑暗处时,你的瞳孔会放大以让更多光线进入,从而使其看物体更清楚。 大多数相机上的闪光灯都非常靠近镜头。当人打开闪光灯拍照时,闪光灯发出的光会通过放大的人的瞳孔到达眼球后部,然后通过瞳孔返回到相机镜头。眼球的后部称为眼底。它是红色的,因为眼底有充足的血液供应。

现在大多数相机闪光灯都会闪烁几秒,这会使瞳孔收缩,从而减少了红眼的可能性。

如何自动去除红眼?

去除红眼的具体操作步骤如下。

1)眼睛检测

去除红眼的第一步是自动检测眼睛。可以使用标准的 OpenCV Haar 检测器来寻找眼睛。有时先运行面部检测器然后检测面部区域内的眼睛是有意义的。为了简单起见,可以直接在图像上运行眼睛检测器。当输入的图像是人像照片或有眼睛特写时,跳过面部检测器会起作用。

Python 代码如下:

```
# Read image
img = cv2.imread("red_eyes.jpg", cv2.IMREAD_COLOR)

# Output image
imgOut = img.copy()

# Load HAAR cascade
eyesCascade = cv2.CascadeClassifier("haarcascade_eye.xml")

# Detect eyes
eyes = eyesCascade.detectMultiScale(img,scaleFactor=1.3, minNeighbors=4, minSize=
(100, 100))
```

2）遮盖红眼

在这一步，需要找到瞳孔中受红眼影响的部分。有许多不同的方法可以找到红色的东西。需要注意的一点是，我们要找的颜色不仅仅是红色，它是鲜红色！可以根据色调和亮度将图像转换为 HSV 颜色空间及阈值。这里使用了一个更简单的启发式方法——红色通道应该大于一个阈值，也应该大于绿色和蓝色通道的总和。出于概念验证系统的目的，启发式方法就足够了，但是，如果想为商业软件包构建自动红眼消除功能，则需要收集数千张红眼图像才能提出更好的方案。

在下面的代码中，首先循环遍历在步骤 2）中检测到的所有眼睛矩形，然后使用 split 命令将彩色图像拆分为 3 个通道，最后为红色通道高于阈值（150）且红色通道大于绿色和蓝色通道之和的所有像素都创建一个掩码，该掩码为 1。

Python 代码如下：

```python
for (x, y, w, h) in eyes:

    # Extract eye from the image
    eye = img[y:y+h, x:x+w]

    # Split eye image into 3 channels
    b = eye[:, :, 0]
    g = eye[:, :, 1]
    r = eye[:, :, 2]

    # Add the green and blue channels
    bg = cv2.add(b, g)

    # Simple red eye detector
    mask = (r > 150) & (r > bg)

    # Convert the mask to uint8 format
    mask = mask.astype(np.uint8)*255
```

3）清理瞳孔面罩

在步骤 2）中创建的蒙版很可能有孔。使用下面的代码移除蒙版中的孔。

```python
def fillHoles(mask):
    maskFloodfill = mask.copy()
    h, w = maskFloodfill.shape[:2]
    maskTemp = np.zeros((h+2, w+2), np.uint8)
    cv2.floodFill(maskFloodfill, maskTemp, (0, 0), 255)
    mask2 = cv2.bitwise_not(maskFloodfill)
    return mask2 | mask
```

此外，扩张蒙版是个好主意，但它所覆盖的区域比必要的稍大。这是因为在边界处颜色逐渐消失，并且原始蒙版中可能没有捕捉到一些红色。使用下面的代码生成扩张蒙版。

```python
# Clean up mask by filling holes and dilating
mask = fillHoles(mask)
mask = cv2.dilate(mask, None, anchor=(-1, -1), iterations=3, borderType=1, borderValue=1)
```

4）修复红眼

现在已有了一个只包含每只眼睛的红色区域的蒙版。接下来将展示如何处理此蒙版内的区域以修复红眼。

由于红眼使图像中的红色通道饱和，换句话说，红色通道中的所有信息都被破坏了。那么，如何才能恢复其中的一些信息？在修复红眼时，不需要在红色通道中检索真正的底层纹理，而只需要找到一个合理的纹理。

幸运的是，红眼效果只破坏了红色通道中的纹理，而蓝色和绿色通道还是不错的。

绿色和蓝色通道的组合可用于得出合理的红色通道。例如，可以创建一个红色通道，它是图像中绿色和蓝色通道的平均值。然而，这样做会给瞳孔带来轻微的淡色，这可能看起来不错，但实际上不是很好。

这又带来了一个重要的问题：瞳孔应该是什么颜色？瞳孔是眼睛的开口，眼睛内部是完全黑暗的。因此，瞳孔应该是无色和深色的。所以，不应该只替换瞳孔区域中的红色通道，而是应该用绿色和蓝色通道的平均值替换所有通道。

下面的代码首先通过平均绿色和蓝色通道来创建平均通道，然后用这个平均通道替换所有通道的掩码区域内的所有像素。

```python
# Calculate the mean channel by averaging
# the green and blue channels. Recall, bg = cv2.add(b, g)
mean = bg / 2
mask = mask.astype(np.bool)[:, :, np.newaxis]
mean = mean[:, :, np.newaxis]

# Copy the eye from the original image
eyeOut = eye.copy()

# Copy the mean image to the output image
np.copyto(eyeOut, mean, where=mask)
```

5）替换固定的眼睛区域

在步骤4）中，已经修复了3个通道。这一步是合并3个通道以创建RGB图像，然后将这个固定的眼睛区域放回原始图像中。

Python 代码如下：

```python
# Copy the fixed eye to the output image
imgOut[y:y+h, x:x+w, :] = eyeOut
```

1.38 创建虚拟笔和橡皮擦

如果可以在空中挥动一支笔以虚拟地绘制一些东西并且实际上将它绘制在屏幕上，那不是很酷吗？如果不使用任何特殊的硬件来实际实现这一点，可能更有趣，而且只是通过简单的计算机视觉就可以做到。

为了实现这一点，首先使用颜色遮罩来获得目标彩色笔的二进制遮罩（这里使用蓝色标记作为虚拟笔），然后使用轮廓检测来检测并跟踪该笔在整个屏幕上的位置。一旦完成了，只需将点从字面上连接起来，即只需使用笔的先前位置（前一帧中的位置）的 x, y 坐标与新的 x, y

点（x，y 点在新帧中）之间画一条线，这样就有了一支虚拟笔。

现在要完成预处理和添加一些其他功能，这里是应用程序中每个步骤的细分。

第 1 步：找到目标对象的颜色范围并保存。

第 2 步：应用正确的形态学运算来减少视频中的噪声。

第 3 步：通过轮廓检测来检测和跟踪有色物体。

第 4 步：找到要在屏幕上绘制的对象的 x、y 位置坐标。

第 5 步：添加擦拭器功能以擦除整个屏幕。

第 6 步：添加橡皮擦功能以擦除部分绘图。

请注意，在上述的第 4 步已准备好了虚拟笔，因此在第 5、6 步中需要添加更多功能。例如，在第 5 步，添加一个虚拟擦拭器，并使其可以像笔一样从屏幕上擦除笔迹；然后在第 6 步，添加橡皮擦功能，并允许使用橡皮擦切换笔。

下面详细介绍上述细分步骤的具体实现。

首先导入所需的库：

```
import cv2
import numpy as np
import time
```

1）找到目标笔的颜色范围并保存

首先必须为目标颜色对象找到一个合适的颜色范围，这个范围将在 cv2.inrange() 函数中使用，以过滤出所需的颜色对象。还需要将范围数组保存为 .npy 文件，以便以后可以访问它。

由于正在尝试进行颜色检测，因此会将 RGB 格式图像转换为 HSV 颜色格式，因为在该模型中操作颜色更容易。

下面的脚本将允许使用轨迹栏来调整图像的色调、饱和度和值通道，直到只有目标对象可见，其余为黑色。

```
# A required callback method that goes into the trackbar function
def nothing(x):
    pass

# Initializing the webcam feed
cap = cv2.VideoCapture(0)
cap.set(3,1280)
cap.set(4,720)

# Create a window named trackbars
cv2.namedWindow("Trackbars")

# Now create 6 trackbars that will control the lower and upper range of
# H,S and V channels. The Arguments are like this: Name of trackbar,
# window name, range,callback function. For Hue the range is 0-179 and
# for S,V its 0-255.
cv2.createTrackbar("L - H", "Trackbars", 0, 179, nothing)
cv2.createTrackbar("L - S", "Trackbars", 0, 255, nothing)
cv2.createTrackbar("L - V", "Trackbars", 0, 255, nothing)
cv2.createTrackbar("U - H", "Trackbars", 179, 179, nothing)
cv2.createTrackbar("U - S", "Trackbars", 255, 255, nothing)
```

```python
cv2.createTrackbar("U - V", "Trackbars", 255, 255, nothing)


while True:
    
    # Start reading the webcam feed frame by frame
    ret, frame = cap.read()
    if not ret:
        break
    # Flip the frame horizontally (Not required)
    frame = cv2.flip( frame, 1 )
    
    # Convert the BGR image to HSV image
    hsv = cv2.cvtColor(frame, cv2.COLOR_BGR2HSV)
    
    # Get the new values of the trackbar in real time as the user changes
    # them
    l_h = cv2.getTrackbarPos("L - H", "Trackbars")
    l_s = cv2.getTrackbarPos("L - S", "Trackbars")
    l_v = cv2.getTrackbarPos("L - V", "Trackbars")
    u_h = cv2.getTrackbarPos("U - H", "Trackbars")
    u_s = cv2.getTrackbarPos("U - S", "Trackbars")
    u_v = cv2.getTrackbarPos("U - V", "Trackbars")
    
    # Set the lower and upper HSV range according to the value selected
    # by the trackbar
    lower_range = np.array([l_h, l_s, l_v])
    upper_range = np.array([u_h, u_s, u_v])
    
    # Filter the image and get the binary mask, where white represents
    # your target color
    mask = cv2.inRange(hsv, lower_range, upper_range)
    
    # You can also visualize the real part of the target color (Optional)
    res = cv2.bitwise_and(frame, frame, mask=mask)
    
    # Converting the binary mask to 3 channel image, this is just so
    # we can stack it with the others
    mask_3 = cv2.cvtColor(mask, cv2.COLOR_GRAY2BGR)
    
    # stack the mask, orginal frame and the filtered result
    stacked = np.hstack((mask_3,frame,res))
    
    # Show this stacked frame at 40% of the size
    cv2.imshow('Trackbars',cv2.resize(stacked,None,fx=0.4,fy=0.4))
    
    # If the user presses Esc then exit the program
    key = cv2.waitKey(1)
    if key == 27:
        break
```

```python
    # If the user presses `s` then print this array
    if key == ord('s'):

        thearray = [[l_h,l_s,l_v],[u_h, u_s, u_v]]
        print(thearray)

        # Also save this array as penval.npy
        np.save('penval',thearray)
        break

# Release the camera & destroy the windows
cap.release()
cv2.destroyAllWindows()
```

2）最大化检测蒙版并消除噪声

现在不需要由步骤 1）得到一个完美的蒙版，图像中有一些像白点这样的噪点是可以的，并且在这一步可以通过形态学操作来去除这些噪点。

现在我使用一个名为 load_from_disk 的变量来决定是要从磁盘加载颜色范围还是要使用一些自定义值。实现代码如下：

```python
# This variable determines if we want to load color range from memory
# or use the ones defined here
load_from_disk = True

# If true then load color range from memory
if load_from_disk:
    penval = np.load('penval.npy')

cap = cv2.VideoCapture(0)
cap.set(3,1280)
cap.set(4,720)

# Creating A 5x5 kernel for morphological operations
kernel = np.ones((5,5),np.uint8)

while(1):

    ret, frame = cap.read()
    if not ret:
        break

    frame = cv2.flip( frame, 1 )

    # Convert BGR to HSV
    hsv = cv2.cvtColor(frame, cv2.COLOR_BGR2HSV)

    # If you're reading from memory then load the upper and lower ranges
    # from there
    if load_from_disk:
```

```
        lower_range = penval[0]
        upper_range = penval[1]

    # Otherwise define your own custom values for upper and lower range.
    else:
        lower_range  = np.array([26,80,147])
        upper_range = np.array([81,255,255])

    mask = cv2.inRange(hsv, lower_range, upper_range)

    # Perform the morphological operations to get rid of the noise
    # Erosion Eats away the white part while dilation expands it
    mask = cv2.erode(mask,kernel,iterations = 1)
    mask = cv2.dilate(mask,kernel,iterations = 2)

    res = cv2.bitwise_and(frame,frame, mask= mask)

    mask_3 = cv2.cvtColor(mask, cv2.COLOR_GRAY2BGR)

    # stack all frames and show it
    stacked = np.hstack((mask_3,frame,res))
    cv2.imshow('Trackbars',cv2.resize(stacked,None,fx=0.4,fy=0.4))

    k = cv2.waitKey(5) & 0xFF
    if k == 27:
        break

cv2.destroyAllWindows()
cap.release()
```

在上面的脚本中，已经执行了 1 次侵蚀迭代和 2 次膨胀迭代，全部使用 5×5 内核。现在重要的是要注意，这些迭代次数和内核大小是特定于目标对象和实验场景照明条件的。对于其他情况来说，这些值可能有用，也可能不适合，最好根据实际情况调整这些值，以便得到最好的结果。

现在首先执行腐蚀以去除那些小白点，然后进行膨胀以扩大目标对象。

即使仍然有一些白噪声也没关系，在下一步可以避免这些白噪声，但要确保目标对象的蒙版在理想情况下清晰可见，并且内部没有任何孔。

3）跟踪目标笔

现在已经有了一个不错的蒙版，可以使用轮廓检测来检测笔了。其方法是在对象周围绘制一个边界框，以确保在整个屏幕上都能检测到它。实现代码如下：

```
# This variable determines if we want to load color range from memory
# or use the ones defined in the notebook
load_from_disk = True

# If true then load color range from memory
if load_from_disk:
    penval = np.load('penval.npy')
```

```python
cap = cv2.VideoCapture(0)
cap.set(3,1280)
cap.set(4,720)

# kernel for morphological operations
kernel = np.ones((5,5),np.uint8)

# set the window to auto-size so we can view this full screen
cv2.namedWindow('image', cv2.WINDOW_NORMAL)

# This threshold is used to filter noise, the contour area must be
# bigger than this to qualify as an actual contour
noiseth = 500

while(1):
    
    _, frame = cap.read()
    frame = cv2.flip( frame, 1 )
    
    # Convert BGR to HSV
    hsv = cv2.cvtColor(frame, cv2.COLOR_BGR2HSV)
    
    # If you're reading from memory then load the upper and lower
    # ranges from there
    if load_from_disk:
        lower_range = penval[0]
        upper_range = penval[1]
    
    # Otherwise define your own custom values for upper and lower range
    else:
        lower_range  = np.array([26,80,147])
        upper_range = np.array([81,255,255])
    
    mask = cv2.inRange(hsv, lower_range, upper_range)
    
    # Perform the morphological operations to get rid of the noise
    mask = cv2.erode(mask,kernel,iterations = 1)
    mask = cv2.dilate(mask,kernel,iterations = 2)
    
    # Find Contours in the frame
    contours, hierarchy = cv2.findContours(mask, cv2.RETR_EXTERNAL,
                             cv2.CHAIN_APPROX_SIMPLE)
    
    # Make sure there is a contour present and also make sure its size
    # is bigger than noise threshold
    if contours and cv2.contourArea(max(contours,
                    key = cv2.contourArea)) > noiseth:
                    
        # Grab the biggest contour with respect to area
        c = max(contours, key = cv2.contourArea)
```

```
        # Get bounding box coordinates around that contour
        x,y,w,h = cv2.boundingRect(c)

        # Draw that bounding box
        cv2.rectangle(frame,(x,y),(x+w,y+h),(0,25,255),2)

    cv2.imshow('image',frame)

    k = cv2.waitKey(5) & 0xFF
    if k == 27:
        break

cv2.destroyAllWindows()
cap.release()
```

4）用笔绘图

现在一切准备就绪，可以轻松地跟踪目标对象了。是时候使用这个对象在屏幕上进行虚拟绘图了。

现在只需要使用从前一帧（F-1）的 cv2.boundingRect()函数返回的 *x*、*y* 位置，并将其与新帧（F）中对象的 *x*、*y* 坐标连接起来。通过连接这两个点，即在两点间画了一条线，并对网络摄像头的所有帧都这样做，这样就可以看到用笔进行的实时绘图。

注意：绘图是在黑色画布上完成，然后将该画布与帧合并。这是因为在每次迭代中都会得到一个新帧，所以我们不能在实际帧上绘制。

用虚拟笔绘图的实现代码如下：

```
load_from_disk = True
if load_from_disk:
    penval = np.load('penval.npy')

cap = cv2.VideoCapture(0)
cap.set(3,1280)
cap.set(4,720)

kernel = np.ones((5,5),np.uint8)

# Initializing the canvas on which we will draw upon
canvas = None

# Initilize x1,y1 points
x1,y1=0,0

# Threshold for noise
noiseth = 800

while(1):
    _, frame = cap.read()
    frame = cv2.flip( frame, 1 )
```

```python
# Initialize the canvas as a black image of the same size as the frame
if canvas is None:
    canvas = np.zeros_like(frame)

# Convert BGR to HSV
hsv = cv2.cvtColor(frame, cv2.COLOR_BGR2HSV)

# If you're reading from memory then load the upper and lower ranges
# from there
if load_from_disk:
        lower_range = penval[0]
        upper_range = penval[1]

# Otherwise define your own custom values for upper and lower range
else:
   lower_range  = np.array([26,80,147])
   upper_range = np.array([81,255,255])

mask = cv2.inRange(hsv, lower_range, upper_range)

# Perform morphological operations to get rid of the noise
mask = cv2.erode(mask,kernel,iterations = 1)
mask = cv2.dilate(mask,kernel,iterations = 2)

# Find Contours
contours, hierarchy = cv2.findContours(mask, cv2.RETR_EXTERNAL, cv2.CHAIN_APPROX_SIMPLE)

# Make sure there is a contour present and also its size is bigger than
# the noise threshold
if contours and cv2.contourArea(max(contours,
                    key = cv2.contourArea)) > noiseth:

    c = max(contours, key = cv2.contourArea)
    x2,y2,w,h = cv2.boundingRect(c)

    # If there were no previous points then save the detected x2,y2
    # coordinates as x1,y1
    # This is true when we writing for the first time or when writing
    # again when the pen had disappeared from view
    if x1 == 0 and y1 == 0:
        x1,y1= x2,y2

    else:
        # Draw the line on the canvas
        canvas = cv2.line(canvas, (x1,y1),(x2,y2), [255,0,0], 4)

    # After the line is drawn the new points become the previous points
    x1,y1= x2,y2
```

```
    else:
        # If there were no contours detected then make x1,y1 = 0
        x1,y1 =0,0

    # Merge the canvas and the frame
    frame = cv2.add(frame,canvas)

    # Optionally stack both frames and show it
    stacked = np.hstack((canvas,frame))
    cv2.imshow('Trackbars',cv2.resize(stacked,None,fx=0.6,fy=0.6))

    k = cv2.waitKey(1) & 0xFF
    if k == 27:
        break

    # When c is pressed clear the canvas
    if k == ord('c'):
        canvas = None

cv2.destroyAllWindows()
cap.release()
```

5）添加图像擦拭器

在上面的脚本中，已创建了一支虚拟笔，当用户在键盘上按下 C 键时，还可清除或擦除屏幕。现在来实现自动化擦除这部分。一个简单的方法是检测目标物体何时离相机太近，如果太近，就清除屏幕。轮廓的大小随着目标物体靠近相机的远近而增减，因此可以通过监控轮廓的大小来实现这一点。

下面要做的另一件事是，警告用户将在几秒内清除屏幕，以便用户可以将对象从帧中取出。实现代码如下：

```
load_from_disk = True
if load_from_disk:
    penval = np.load('penval.npy')

cap = cv2.VideoCapture(0)
cap.set(3,1280)
cap.set(4,720)

kernel = np.ones((5,5),np.uint8)

# Making window size adjustable
cv2.namedWindow('image', cv2.WINDOW_NORMAL)

# This is the canvas on which we will draw upon
canvas=None

# Initilize x1,y1 points
x1,y1=0,0

# Threshold for noise
```

```python
noiseth = 800

# Threshold for wiper, the size of the contour must be bigger than for us to
# clear the canvas
wiper_thresh = 40000

# A variable which tells when to clear canvas, if its True then we clear the canvas
clear = False

while(1):
    _, frame = cap.read()
    frame = cv2.flip( frame, 1 )

    # Initialize the canvas as a black image
    if canvas is None:
        canvas = np.zeros_like(frame)

    # Convert BGR to HSV
    hsv = cv2.cvtColor(frame, cv2.COLOR_BGR2HSV)

    # If you're reading from memory then load the upper and lower ranges
    # from there
    if load_from_disk:
            lower_range = penval[0]
            upper_range = penval[1]

    # Otherwise define your own custom values for upper and lower range
    else:
       lower_range  = np.array([26,80,147])
       upper_range  = np.array([81,255,255])

    mask = cv2.inRange(hsv, lower_range, upper_range)

    # Perform the morphological operations to get rid of the noise
    mask = cv2.erode(mask,kernel,iterations = 1)
    mask = cv2.dilate(mask,kernel,iterations = 2)

    # Find Contours
    contours, hierarchy = cv2.findContours(mask,
    cv2.RETR_EXTERNAL,cv2.CHAIN_APPROX_SIMPLE)

    # Make sure there is a contour present and also its size is bigger than
    # the noise threshold
    if contours and cv2.contourArea(max(contours,
                        key = cv2.contourArea)) > noiseth:

        c = max(contours, key = cv2.contourArea)
        x2,y2,w,h = cv2.boundingRect(c)

        # Get the area of the contour
```

```python
        area = cv2.contourArea(c)

        # If there were no previous points then save the detected x2,y2
        # coordinates as x1,y1
        if x1 == 0 and y1 == 0:
            x1,y1= x2,y2

        else:
            # Draw the line on the canvas
            canvas = cv2.line(canvas, (x1,y1),(x2,y2),
            [255,0,0], 5)

        # After the line is drawn the new points become the previous points
        x1,y1= x2,y2

        # Now if the area is greater than the wiper threshold then set the
        # clear variable to True and warn User
        if area > wiper_thresh:
            cv2.putText(canvas,'Clearing Canvas', (100,200),
            cv2.FONT_HERSHEY_SIMPLEX,2, (0,0,255), 5, cv2.LINE_AA)
            clear = True

    else:
        # If there were no contours detected then make x1,y1 = 0
        x1,y1 =0,0


    # Now this piece of code is just for smooth drawing. (Optional)
    _ , mask = cv2.threshold(cv2.cvtColor(canvas, cv2.COLOR_BGR2GRAY), 20,
    255, cv2.THRESH_BINARY)
    foreground = cv2.bitwise_and(canvas, canvas, mask = mask)
    background = cv2.bitwise_and(frame, frame,
    mask = cv2.bitwise_not(mask))
    frame = cv2.add(foreground,background)

    cv2.imshow('image',frame)

    k = cv2.waitKey(5) & 0xFF
    if k == 27:
        break

    # Clear the canvas after 1 second if the clear variable is true
    if clear == True:

        time.sleep(1)
        canvas = None

        # And then set clear to false
        clear = False
```

```
cv2.destroyAllWindows()
cap.release()
```

6）添加橡皮擦功能

现在已经实现了笔和擦除器功能，是时候添加橡皮擦功能了。这里所要做的是，当用户切换到橡皮擦而不是绘图时，它会擦除用笔所绘制的那部分。这真的很容易——只需要在画布上用橡皮擦画黑色就可以了。通过绘制黑色，该部分在合并期间恢复为原始状态，因此它就像橡皮擦一样。橡皮擦功能的真正编码部分是如何在笔和橡皮擦之间进行切换，当然最简单的方法是使用键盘按键，但是有比这更酷的方法。

下面要做的是，每当有人把手放在屏幕的左上角时就执行切换。可以使用背景减法来监控该区域，以便知道何时有干扰。这就像按下一个虚拟按键。

实现代码如下：

```python
load_from_disk = True
if load_from_disk:
    penval = np.load('penval.npy')

cap = cv2.VideoCapture(0)

# Load these 2 images and resize them to the same size
pen_img = cv2.resize(cv2.imread('pen.png',1), (50, 50))
eraser_img = cv2.resize(cv2.imread('eraser.jpg',1), (50, 50))

kernel = np.ones((5,5),np.uint8)

# Making window size adjustable
cv2.namedWindow('image', cv2.WINDOW_NORMAL)

# This is the canvas on which we will draw upon
canvas = None

# Create a background subtractor Object
backgroundobject = cv2.createBackgroundSubtractorMOG2(detectShadows = False)

# This threshold determines the amount of disruption in the background
background_threshold = 600

# A variable which tells you if you're using a pen or an eraser
switch = 'Pen'

# With this variable we will monitor the time between previous switch
last_switch = time.time()

# Initilize x1,y1 points
x1,y1=0,0

# Threshold for noise
noiseth = 800
```

```python
# Threshold for wiper, the size of the contour must be bigger than this for # us to
# clear the canvas
wiper_thresh = 40000

# A variable which tells when to clear canvas
clear = False

while(1):
    _, frame = cap.read()
    frame = cv2.flip( frame, 1 )

    # Initilize the canvas as a black image
    if canvas is None:
        canvas = np.zeros_like(frame)

    # Take the top left of the frame and apply the background subtractor
    # there
    top_left = frame[0: 50, 0: 50]
    fgmask = backgroundobject.apply(top_left)

    # Note the number of pixels that are white, this is the level of
    # disruption
    switch_thresh = np.sum(fgmask==255)

    # If the disruption is greater than background threshold and there has
    # been some time after the previous switch then you. can change the
    # object type
    if switch_thresh>background_threshold and (time.time()-last_switch) > 1:

        # Save the time of the switch
        last_switch = time.time()

        if switch == 'Pen':
            switch = 'Eraser'
        else:
            switch = 'Pen'

    # Convert BGR to HSV
    hsv = cv2.cvtColor(frame, cv2.COLOR_BGR2HSV)

    # If you're reading from memory then load the upper and lower ranges
    # from there
    if load_from_disk:
            lower_range = penval[0]
            upper_range = penval[1]

    # Otherwise define your own custom values for upper and lower range
    else:
       lower_range  = np.array([26,80,147])
       upper_range  = np.array([81,255,255])
```

```python
mask = cv2.inRange(hsv, lower_range, upper_range)

# Perform morphological operations to get rid of the noise
mask = cv2.erode(mask,kernel,iterations = 1)
mask = cv2.dilate(mask,kernel,iterations = 2)

# Find Contours
contours, _ = cv2.findContours(mask, cv2.RETR_EXTERNAL,
cv2.CHAIN_APPROX_SIMPLE)

# Make sure there is a contour present and also it size is bigger than
# noise threshold
if contours and cv2.contourArea(max(contours,
                        key = cv2.contourArea)) > noiseth:

    c = max(contours, key = cv2.contourArea)
    x2,y2,w,h = cv2.boundingRect(c)

    # Get the area of the contour
    area = cv2.contourArea(c)

    # If there were no previous points then save the detected x2,y2
    # coordinates as x1,y1
    if x1 == 0 and y1 == 0:
        x1,y1= x2,y2

    else:
        if switch == 'Pen':
            # Draw the line on the canvas
            canvas = cv2.line(canvas, (x1,y1),
            (x2,y2), [255,0,0], 5)

        else:
            cv2.circle(canvas, (x2, y2), 20,
            (0,0,0), -1)


    # After the line is drawn the new points become the previous points
    x1,y1= x2,y2

    # Now if the area is greater than the wiper threshold then set the
    # clear variable to True
    if area > wiper_thresh:
        cv2.putText(canvas,'Clearing Canvas',(0,200),
        cv2.FONT_HERSHEY_SIMPLEX, 2, (0,0,255), 1, cv2.LINE_AA)
        clear = True

else:
```

```
        # If there were no contours detected then make x1,y1 = 0
        x1,y1 =0,0

    # Now this piece of code is just for smooth drawing. (Optional)
    _ , mask = cv2.threshold(cv2.cvtColor (canvas, cv2.COLOR_BGR2GRAY), 20,
    255, cv2.THRESH_BINARY)
    foreground = cv2.bitwise_and(canvas, canvas, mask = mask)
    background = cv2.bitwise_and(frame, frame,
    mask = cv2.bitwise_not(mask))
    frame = cv2.add(foreground,background)

    # Switch the images depending upon what we're using, pen or eraser
    if switch != 'Pen':
        cv2.circle(frame, (x1, y1), 20, (255,255,255), -1)
        frame[0: 50, 0: 50] = eraser_img
    else:
        frame[0: 50, 0: 50] = pen_img

    cv2.imshow('image',frame)

    k = cv2.waitKey(5) & 0xFF
    if k == 27:
        break

    # Clear the canvas after 1 second, if the clear variable is true
    if clear == True:
        time.sleep(1)
        canvas = None

        # And then set clear to false
        clear = False

cv2.destroyAllWindows()
cap.release()
```

1.39 使用 ArUco 标记的增强现实

本节将解释什么是 ArUco 标记，以及如何使用 OpenCV 将它们用于简单的增强现实任务。ArUco 标记已在增强现实、相机姿态估计和相机校准中使用了一段时间。

ArUco 标记最初是由 S.Garrido-Jurado 等于 2014 年在他们的论文《遮挡下高可靠基准标记的自动生成和检测》中提出的。

ArUco 标记是放置在被成像对象或场景上的基准标记。它是一个具有黑色背景和边界的二进制正方形，其中有一个由白色生成的图案，可以唯一地被识别。黑色边界有助于使它们的检测更容易。它们可以以各种尺寸生成。根据对象大小和场景选择大小，可以成功检测。如果没有检测到非常小的标记，只须增加它们的大小就可以使检测更容易。

可以打印一些标记并将它们放在现实世界中，也可以拍摄现实世界并独特地检测这些标记。一些初学者会想：这有什么用？看下面的几个用例。

将打印好的标记放在相框的角落。当唯一地识别标记时，能够用任意视频或图像替换相框。当移动相机时，新图片具有正确的透视失真。

在机器人应用程序中，可以将这些标记放置在配备摄像头的仓库机器人的路径上。当安装在机器人上的摄像头检测到这些标记中的一个时，仓库机器人就可以知道这些标记在仓库中的精确位置，因为每个标记都有一个唯一的 ID，所以就可以通过 ID 知道标记所对应的物品在仓库中的放置位置。

可以使用 OpenCV 非常轻松地生成这些标记。OpenCV 中的 ArUco 模块共有 25 个预定义的标记字典。字典中的所有标记包含相同数量的块或位（4×4、5×5、6×6 或 7×7），每个字典包含固定数量的标记（50、100、250 或 1000）。下面展示如何在 Python 中生成和检测各种 ArUco 标记。

首先需要在代码中使用 aruco 模块。

下面的函数调用 getPredefinedDictionary() 显示了如何加载包含 250 个标记的字典，其中每个标记包含一个 6×6 位的二进制模式。

```python
import cv2 as cv
import numpy as np

# Load the predefined dictionary
dictionary = cv.aruco.Dictionary_get(cv.aruco.DICT_6X6_250)

# Generate the marker
markerImage = np.zeros((200, 200), dtype=np.uint8)
markerImage = cv.aruco.drawMarker(dictionary, 33, 200, markerImage, 1);

cv.imwrite("marker33.png", markerImage);
```

上面的 drawMarker() 函数可以从 0～249 的 250 个 ID 标记集合中选择具有给定 ID（第二个参数 33）的标记。drawMarker() 函数的第三个参数决定生成的标记的大小。在上面的例子中，它将生成一个 200×200 像素的图像。第四个参数表示将存储生成的标记（上面的 markerImage）对象。最后，第五个参数是厚度参数，它决定了应该将多少块作为边界添加到生成的二进制模式中。在上面的示例中，将在 6×6 位生成的图案周围添加 1 位的边界，以在 200×200 像素的图像中生成具有 7×7 位的图像。

对于大多数应用程序，需要生成多个标记，然后打印它们并放入场景中。

使用 ArUco 标记对场景进行成像后，需要检测它们并使用它们对图像进行进一步处理。下面展示如何检测标记。

```python
#Load the dictionary that was used to generate the markers.
dictionary = cv.aruco.Dictionary_get(cv.aruco.DICT_6X6_250)

# Initialize the detector parameters using default values
parameters =  cv.aruco.DetectorParameters_create()

# Detect the markers in the image
markerCorners, markerIds, rejectedCandidates = cv.aruco.detectMarkers(frame, dictionary,
```

parameters=parameters)

首先加载用来生成标记的同一个字典。

使用 DetectorParameters_create() 检测一组初始参数。OpenCV 允许在检测过程中更改多个参数。在大多数情况下，默认参数运行良好，OpenCV 建议使用这些参数。

对于每个成功的标记检测，从左上角、右上角、右下角和左下角依次检测标记的 4 个角点。在 Python 中，它们存储为 NumPy 数组。

detectMarkers() 函数用于检测和定位标记的角点。第一个参数是带有标记的场景图像；第二个参数是用于生成标记的字典。成功检测到的标记将存储在 markerCorners 中，它们的 ID 存储在 markerIds 中。之前初始化的 DetectorParameters 对象也作为参数传递。最后，被拒绝的候选标记存储在 rejectedCandidates 中。

在场景中打印、切割和放置标记时，重要的是在标记的黑色边界周围保留一些白色边框，以便可以轻松地检测到它们。

ArUco 标记主要是为解决包括增强现实在内的各种应用的相机姿态估计问题而开发的。

本节会将它们用于增强现实应用程序，以便可以将任何新场景叠加到现有图像或视频上。现在挑选一个带有大相框的场景，以用新的图片替换框架中的图片，看看它在墙上的样子，然后继续尝试在影片中插入视频。为此，在图像区域的角落打印、剪切和粘贴大的 ArUco 标记，然后捕获视频。

输入图像和新场景图像中的 4 个对应点集用于计算单应性。给定在场景不同视图中的对应点，单应性是将一个对应点映射到另一个对应点的变换。

在下面的例子中，单应矩阵用于将新的场景图像变形为由所捕获的图像中的标记定义的四边形。下面的代码展示了如何做到这一点。

```python
# Calculate Homography
h, status = cv.findHomography(pts_src, pts_dst)

# Warp source image to destination based on homography
warped_image = cv.warpPerspective(im_src, h, (frame.shape[1],frame.shape[0]))

# Prepare a mask representing region to copy from the warped image into the original
# frame
mask = np.zeros([frame.shape[0], frame.shape[1]], dtype=np.uint8);
cv.fillConvexPoly(mask, np.int32([pts_dst_m]), (255, 255, 255), cv.LINE_AA);

# Erode the mask to not copy the boundary effects from the warping
element = cv.getStructuringElement(cv.MORPH_RECT, (3,3));
mask = cv.erode(mask, element, iterations=3);

# Copy the mask into 3 channels
warped_image = warped_image.astype(float)
mask3 = np.zeros_like(warped_image)
for i in range(0, 3):
    mask3[:,:,i] = mask/255

# Copy the masked warped image into the original frame in the mask region
warped_image_masked = cv.multiply(warped_image, mask3)
```

```
frame_masked = cv.multiply(frame.astype(float), 1-mask3)
im_out = cv.add(warped_image_masked, frame_masked)
```

之后使用新的场景图像角点作为源点（pts_src），并将所捕获的图像中相框内的相应图片角点作为目标点（dst_src）。OpenCV 的 findHomography()函数计算源点和目标点之间的单应矩阵形 h。然后使用单应矩阵来扭曲新图像以适合目标帧。扭曲的图像被屏蔽并复制到目标帧中。在视频情况下，这个过程在每帧上重复。

第 2 章
特征检测和描述

还记得小时候玩的拼图游戏吗？目的是将拼图拼凑在一起。当拼图全部拼好后，就能看到全局，通常是某个人、某个地方、某个事物或三者的组合。

是什么让人们成功完成了这个谜题？每个拼图都包含一些线索，它们可能是边缘、角点、特定颜色或图案等。计算机视觉中的一个特征是图像中唯一且易于识别的感兴趣区域。特征包括点、边、斑点和角点等。

角点是图像中在所有方向上强度变化很大的区域。本章首先介绍 Harris 角点检测算法。

本章将讲解什么是特征，为什么它们很重要，为什么角点很重要等问题。

2.1 Harris 角点检测

OpenCV 的函数 cv.cornerHarris() 实现了 Harris 角点检测算法。其参数有如下几个。
- Img：输入图像。它应该是灰度图和 float32 类型。
- blockSize：角点检测考虑的邻域大小。
- ksize：使用的 Sobel 导数的孔径参数。
- k：方程中的 Harris 检测器自由参数。

请参见下面的代码：

```
import numpy as np
import cv2 as cv
filename = 'chessboard.png'
img = cv.imread(filename)
gray = cv.cvtColor(img,cv.COLOR_BGR2GRAY)
gray = np.float32(gray)
dst = cv.cornerHarris(gray,2,3,0.04)
#result is dilated for marking the corners, not important
dst = cv.dilate(dst,None)
# Threshold for an optimal value, it may vary depending on the image
img[dst>0.01*dst.max()]=[0,0,255]
```

```
cv.imshow('dst',img)
if cv.waitKey(0) & 0xff == 27:
    cv.destroyAllWindows()
```

2.2 Shi-Tomasi 角点检测器和良好的跟踪功能

OpenCV 有一个 cv.goodFeaturesToTrack()函数。它通过 Shi-Tomasi 方法在图像中找到 N 个最强角点。像往常一样，图像应该是灰度图像。然后指定要查找的角点数，并指定质量级别。质量级别是一个介于 0 和 1 之间的值，表示每个角落都被拒绝的最低拐角质量。最后提供检测到的角点之间的最小欧几里得距离。

有了所有这些信息，该函数就可以找到图像中的角点。所有低于质量水平的角点都会被忽略。它可以根据质量按降序对剩余的角点进行排序。然后函数取第一个最佳角点，扔掉最小距离范围内的所有附近角点，并返回 N 个最佳角点。

在下面的代码中，将尝试找到 25 个最佳角点：

```
import numpy as np
import cv2 as cv
from matplotlib import pyplot as plt
img = cv.imread('blox.jpg')
gray = cv.cvtColor(img,cv.COLOR_BGR2GRAY)
corners = cv.goodFeaturesToTrack(gray,25,0.01,10)
corners = np.int0(corners)
for i in corners:
    x,y = i.ravel()
    cv.circle(img,(x,y),3,255,-1)
plt.imshow(img),plt.show()
```

2.3 尺度不变特征变换

当图像比例发生变化时，Harris 角点检测器的性能不理想。D.Lowe 开发了一种突破性的方法来寻找尺度不变的特征，它被称为尺度不变特征变换（Scale-Invariant Feature Transform，SIFT）。

下面从关键点检测开始并绘制它们。首先，必须构造一个 SIFT 对象，并可以向它传递不同的参数，这些参数是可选的。实现代码如下：

```
import numpy as np
import cv2 as cv
img = cv.imread('home.jpg')
gray= cv.cvtColor(img,cv.COLOR_BGR2GRAY)
sift = cv.SIFT_create()
kp = sift.detect(gray,None)
img=cv.drawKeypoints(gray,kp,img)
cv.imwrite('sift_keypoints.jpg',img)
```

利用 sift.detect()函数可以找到图像中的关键点。如果只想搜索图像的一部分，则可以传递掩码。每个关键点都是一个特殊的结构，它具有许多属性，例如它的 (x, y) 坐标、有意义的邻域的大小、指定的方向的角度、指定的关键点强度的响应等。

OpenCV 还提供了 cv.drawKeyPoints()函数，用于在关键点的位置上绘制小圆圈。如果将标志 cv.DRAW_MATCHES_FLAGS_DRAW_RICH_KEYPOINTS 传递给它，它将绘制一个关键点大小的圆，甚至会显示其方向。请参见下面的示例代码。

```
img=cv.drawKeypoints(gray,kp,img,flags=cv.DRAW_MATCHES_FLAGS_DRAW_RICH_KEYPOINTS)
cv.imwrite('sift_keypoints.jpg',img)
```

现在要计算描述符，对此 OpenCV 提供了以下两种方法。

（1）如果已经找到了关键点，就可以调用 sift.compute()函数从找到的关键点计算描述符。例如：

```
kp,des = sift.compute(gray,kp)
```

（2）如果没有找到关键点，则可直接使用 sift.detectAndCompute()函数一步找到关键点和描述符。

来看看第二种方法：

```
sift = cv.SIFT_create()
kp, des = sift.detectAndCompute(gray,None)
```

这里的 kp 将是一个关键点列表，而 des 是一个形状为关键点数×128 的 NumPy 数组。

现在已得到了关键点、描述符等。接下来看看如何匹配不同图像中的关键点。

2.4 特征匹配

本节将讲解如何在 OpenCV 中使用 Brute-Force 匹配器和 FLANN 匹配器将一张图像中的特征与其他图像中的特征进行匹配。

1）Brute-Force（强力）匹配器

Brute-Force（强力）匹配器很简单。它采用第一个集合中的一个特征的描述符，并使用一些距离计算与第二个集合中的所有其他特征匹配，并返回最接近的那个。

对于 Brute-Force 匹配器，首先必须使用 cv.BFMatcher()创建 BFMatcher 对象。它需要两个可选参数。

第一个参数是 normType。它指定要使用的距离度量。默认情况下，它是 cv.NORM_L2。它适用于 SIFT、SURF 等。对于基于二进制字符串的描述符，如 ORB、BRIEF、BRISK 等，应使用 cv.NORM_HAMMING，它使用汉明距离作为度量。如果 ORB 使用 WTA_K == 3 或 4，则应使用 cv.NORM_HAMMING2。

第二个参数是布尔变量 crossCheck，默认为 false。如果为 true，则匹配器仅返回值为(i, j)的匹配项，使得集合 A 中的第 i 个描述符具有集合 B 中的第 j 个描述符作为最佳匹配，反之亦然。也就是说，两个集合中的两个特征应该相互匹配。

创建 BFMatcher 对象后，两个重要的方法是 BFMatcher.match()和 BFMatcher.knnMatch()。第一个方法返回最佳匹配；第二个方法返回 k 个最佳匹配，其中 k 由用户指定。

就像使用 cv.drawKeypoints()来绘制关键点一样，使用 cv.drawMatches()来绘制匹配。它水平堆叠两幅图像，并从第一幅图像到第二幅图像绘制线条，显示最佳匹配。还有 cv.drawMatchesKnn 用来绘制所有 k 个最佳匹配。如果 k=2，它将为每个关键点绘制两条匹配线。

这一次，将使用 BFMatcher.knnMatch()来获得 k 个最佳匹配。在这个例子中，让 k=2，这样就可以应用比率测试。例子的实现代码如下：

```python
import numpy as np
import cv2 as cv
import matplotlib.pyplot as plt
img1 = cv.imread('box.png',cv.IMREAD_GRAYSCALE)          # queryImage
img2 = cv.imread('box_in_scene.png',cv.IMREAD_GRAYSCALE) # trainImage
# Initiate SIFT detector
sift = cv.SIFT_create()
# find the keypoints and descriptors with SIFT
kp1, des1 = sift.detectAndCompute(img1,None)
kp2, des2 = sift.detectAndCompute(img2,None)
# BFMatcher with default params
bf = cv.BFMatcher()
matches = bf.knnMatch(des1,des2,k=2)
# Apply ratio test
good = []
for m,n in matches:
    if m.distance < 0.75*n.distance:
        good.append([m])
# cv.drawMatchesKnn expects list of lists as matches
img3 = cv.drawMatchesKnn(img1,kp1,img2,kp2,good,None,flags=cv.DrawMatchesFlags_NOT_DRAW_SINGLE_POINTS)
plt.imshow(img3),plt.show()
```

2）FLANN 匹配器

FLANN 匹配器即基于 FLANN（Fast Library for Approximate Nearest Neighbors，快速最近邻搜索包）的匹配器，它包含一组针对大型数据集中的快速最近邻搜索和高维特征进行优化的算法。对于大型数据集，它比 BFMatcher 工作得更快。

对于 FLANN 匹配器，需要传递两个字典，指定要使用的算法、相关参数等。

第一个字典是 IndexParams。对于 SIFT、SURF 等算法，可以传递以下字典：

```python
FLANN_INDEX_KDTREE = 1
index_params = dict(algorithm = FLANN_INDEX_KDTREE, trees = 5)
```

第二个字典是 SearchParams。它指定递归遍历的次数。值越高，精度越高，但也需要更多时间。

有了这些信息，就可以开始进行特征匹配了。实现代码如下：

```python
import numpy as np
import cv2 as cv
import matplotlib.pyplot as plt
img1 = cv.imread('box.png',cv.IMREAD_GRAYSCALE)          # queryImage
img2 = cv.imread('box_in_scene.png',cv.IMREAD_GRAYSCALE) # trainImage
# Initiate SIFT detector
sift = cv.SIFT_create()
# find the keypoints and descriptors with SIFT
kp1, des1 = sift.detectAndCompute(img1,None)
kp2, des2 = sift.detectAndCompute(img2,None)
# FLANN parameters
FLANN_INDEX_KDTREE = 1
```

```
index_params = dict(algorithm = FLANN_INDEX_KDTREE, trees = 5)
search_params = dict(checks=50)   # or pass empty dictionary
flann = cv.FlannBasedMatcher(index_params,search_params)
matches = flann.knnMatch(des1,des2,k=2)
# Need to draw only good matches, so create a mask
matchesMask = [[0,0] for i in range(len(matches))]
# ratio test as per Lowe's paper
for i,(m,n) in enumerate(matches):
    if m.distance < 0.7*n.distance:
        matchesMask[i]=[1,0]
draw_params = dict(matchColor = (0,255,0),
                   singlePointColor = (255,0,0),
                   matchesMask = matchesMask,
                   flags = cv.DrawMatchesFlags_DEFAULT)
img3 = cv.drawMatchesKnn(img1,kp1,img2,kp2,matches,None,**draw_params)
plt.imshow(img3,),plt.show()
```

2.5 特征匹配+单应性查找对象

本节将讲解联合使用特征提取和calib3d模块中的cv.findHomography()函数在复杂图像中查找已知对象。

2.4节使用一个图像queryImage，在其中找到了一些特征点，使用另一个图像trainImage，也找到了该图像中的特征点，并在其中找到了最佳匹配。简而言之，在另一个杂乱的图像中找到了对象某些部分的位置。这些信息足以准确地在图像trainImage上找到对象。

为此，可以使用来自 calib3d 模块的 cv.findHomography()函数。如果从两个图像中传递一组点，它将找到该对象的透视变换。然后我们可以使用 cv.perspectiveTransform()函数来查找对象。它需要至少 4 个正确的点才能找到变换。

我们已经看到，匹配时会出现一些可能影响结果的错误。为了解决这个问题，算法使用RANSAC 或 LEAST_MEDIAN（可以由标志决定）。因此，提供正确估计的良好匹配称为内部点，其余的称为外部点。cv.findHomography()函数返回一个指定内部点和外部点的掩码。

首先，在图像中找到 SIFT 特征并应用比率测试来找到最佳匹配。实现代码如下：

```
import numpy as np
import cv2 as cv
from matplotlib import pyplot as plt
MIN_MATCH_COUNT = 10
img1 = cv.imread('box.png',0)          # queryImage
img2 = cv.imread('box_in_scene.png',0) # trainImage
# Initiate SIFT detector
sift = cv.SIFT_create()
# find the keypoints and descriptors with SIFT
kp1, des1 = sift.detectAndCompute(img1,None)
kp2, des2 = sift.detectAndCompute(img2,None)
FLANN_INDEX_KDTREE = 1
index_params = dict(algorithm = FLANN_INDEX_KDTREE, trees = 5)
search_params = dict(checks = 50)
```

```
flann = cv.FlannBasedMatcher(index_params, search_params)
matches = flann.knnMatch(des1,des2,k=2)
# store all the good matches as per Lowe's ratio test
good = []
for m,n in matches:
    if m.distance < 0.7*n.distance:
        good.append(m)
```

其次,设置一个条件,即至少有 10 个匹配项(由 MIN_MATCH_COUNT 定义)可以找到对象;否则,只需显示一条消息,说明没有足够的匹配项。

如果找到足够的匹配项,我们会提取两个图像中匹配关键点的位置,并将它们传递给 cv.findHomography()函数以找到透视变换。一旦我们得到这个 3×3 的变换矩阵,我们就用它把图像 queryImage 中的角点变换到图像 trainImage 中的对应点。然后我们将其画出来。实现代码如下:

```
if len(good)>MIN_MATCH_COUNT:
    src_pts = np.float32([ kp1[m.queryIdx].pt for m in good ]).reshape(-1,1,2)
    dst_pts = np.float32([ kp2[m.trainIdx].pt for m in good ]).reshape(-1,1,2)
    M, mask = cv.findHomography(src_pts, dst_pts, cv.RANSAC,5.0)
    matchesMask = mask.ravel().tolist()
    h,w = img1.shape
    pts = np.float32([ [0,0],[0,h-1],[w-1,h-1],[w-1,0] ]).reshape(-1,1,2)
    dst = cv.perspectiveTransform(pts,M)
    img2 = cv.polylines(img2,[np.int32(dst)],True,255,3, cv.LINE_AA)
else:
    print( "Not enough matches are found - {}/{}".format(len(good), MIN_MATCH_COUNT) )
    matchesMask = None
```

最后,绘制内点(如果成功找到对象)或匹配关键点(如果失败)。实现代码如下:

```
draw_params = dict(matchColor = (0,255,0), # draw matches in green color
                   singlePointColor = None,
                   matchesMask = matchesMask, # draw only inliers
                   flags = 2)
img3 = cv.drawMatches(img1,kp1,img2,kp2,good,None,**draw_params)
plt.imshow(img3, 'gray'),plt.show()
```

第 3 章
OCR 文字识别

光学字符识别（Optical Character Recognition，OCR）是指用电子设备对文本资料的图像文件进行分析、识别处理，获取文字及版面信息的过程。

3.1 OpenCV 的 OCR 功能

OpenCV 的 OCR 功能旨在读取用户提供的图像文件，然后识别图像中给出的文本以显示给用户。该文本可以进一步用于用户可能需要使用提取的文本的目的。

OCR 用户提供的图像首先需要改变其颜色空间并将其作为变量进行临时存储。OpenCV 的 cvtColor()函数用于颜色空间的转换。函数的第三个参数为颜色空间转换的标识符，用于确定图像发生的转换类型。用户可以根据图像的色彩饱和度和颜色空间的比例，选择灰度缩放或 HSV（Hue，Saturation，Value）颜色模型来转换图像中的红色、绿色和蓝色。

借助 OpenCV 的阈值函数可对图像应用阈值化。有三种阈值化方法可以应用于结果图像：简单阈值、自适应阈值和 Otsu 阈值或二值化方法。

为了从阈值化后的图像提取矩形结构，OpenCV 的 getStructuringElement()函数被用于定义结构元素，如圆形、椭圆形或矩形。在这里，使用矩形结构元素（cv2.MORPH_RECT）。为了制作更大的块并将文本添加到一起，可能需要更大的内核大小。选择内核后，使用 OpenCV 的 dilate()函数将膨胀方法应用于图像，得到膨胀的图像这使得文本块内的文本检测更加精确。

在此之后，需要对用户提供的图像执行的下一个功能是查找轮廓。使用 OpenCV 的查找轮廓功能，可从膨胀的图像中提供返回的轮廓和层次结构。然后，从源图像导出的每个轮廓都保存在一个 NumPy 数组中，其中的坐标对应于图像中存在的对象的边界点。

让轮廓功能用于找到图像中存在的白色物体并取出对比背景，这将是黑色或黑色背景。这种轮廓绘制过程能够检测图像中存在的给定文本块的边界边缘。系统打开一个文本文件以写入提取的文本，然后刷新，在另一端保存执行 OCR 功能提取的文本。

最后，出现了 OCR 的应用。该函数在每个轮廓绘制过程中循环，使用 OpenCV 的边界矩形函数对指定图像的坐标以及高度和宽度进行处理。然后，系统在输出图像中使用 OpenCV 的矩形函数利用获得的坐标以及高度和宽度绘制一个矩形。

Tesseract 是一种开源文本识别引擎，在 Apache 2.0 许可下可用，其开发自 2006 年以来一直由 Google 赞助。可以直接使用它，也可以使用 API 从图像中提取打印文本。通过包装器，Tesseract 可以与不同的编程语言和框架兼容。在这里，将使用名为 pytesseract 的 Python 包装器。它用于从大型文档中识别文本，也可用于从单个文本行的图像中识别文本。

OCR 代码如下：

```python
# import required packages for performing OCR
import cv2
import pytesseract
pytesseract.pytesseract.tesseract_cmd = 'System_path_to_tesseract.exe'
# Reading image file from where the text is to be extracted
img1 = cv2.imread("EduCBA logo.jpg")
# Converting the image into to gray scaled image
Gray1 = cv2.cvtColor(img1, cv2.COLOR_BGR2GRAY)
ret1, thresh_1 = cv2.threshold(gray1, 0, 255, cv2.THRESH_OTSU | cv2.THRESH_BINARY_INV)
# specifying structure shape, kernel size, increase/decreases the kernel area
rect_kernel1 = cv2.getStructuringElement(cv2.MORPH_RECT, (18, 18))
dilation1 = cv2.dilate(thresh1, rect_kernel1, iterations = 1)
# finding contouring for the image
contours1, hierarchy1 = cv2.findcontours(dilation1, cv2.RETR_EXTERNAL,cv2.CHAIN_APPROX_NONE)
# creating a copy
img2 = img1.copy()
file1 = open("recognized.txt", "w+")
file.write("")
file.close()
# looping for ocr through the contours found
for cnt in contours:
    x1, y1, w1, h1 = cv2.boundingrect(cnt)
    rect1 = cv2.rectangle(img2, (x1, y1), (x1 + w1, y1 + h1), (0, 255, 0), 2)
    cropped1 = img2[y1:y1 + h1, x1:x1 + w1]
    file_1 = open("recognized.txt", "a")
    # apply OCR
    text_1 = pytesseract.image_to_string(cropped1)
    file.write(text1)
    file.close
```

Tesseract 4.00 在其新的神经网络子系统中配置了一个文本行识别器。如今，人们通常使用卷积神经网络（Convolutional Neural Networks，CNN）来识别包含单个字符的图像。使用循环神经网络（Recurrent Neural Network，RNN）和长短期记忆（Long-Short Term Memory，LSTM）神经网络解决具有任意长度和字符序列的文本，其中 LSTM 是一种流行的 RNN 形式。

默认情况下，Tesseract 将输入图像视为分段中的一页文本。如果有兴趣从图像中捕获小部分文本，则可以配置 Tesseract 的不同分割。可以通过为其分配 --psm 模式来实现。Tesseract 可实现完全自动化页面分割，但它不执行方向和脚本检测。下面提到了 Tesseract 的不同配置参数：

Page Segmentation Mode（--psm）：通过配置它，可以帮助 Tesseract 以文本形式分割图像。

命令行帮助有 13 种模式。可以从表 3-1 中选择最适合要求的一种。

表 3-1 页面分割模式

模　式	描　述
0	仅限方向和脚本检测（OSD）
1	带 OSD 的自动页面分割
2	自动页面分割，但没有 OSD 或 OCR
3	全自动页面分割，但没有 OSD（默认）
4	假设一列可变大小的文本
5	假设一个具有垂直对齐文本的统一块
6	假设一个统一的文本块
7	将图像视为单个文本行
8	将图像视为单个单词
9	将图像视为圆圈中的单个单词
10	将图像视为单个字符
11	稀疏文本。尽可能多地查找不按特定顺序排列的文本
12	带有 OSD 的稀疏文本
13	原始线

请下载 Tesseract 引擎可执行文件（.exe）并将其安装在 "C:\Program Files\Tesseract-OCR\" 目录中。不要忘记配置 path 环境变量并添加 Tesseract 路径。

安装 pytesseract 模块：

```
>pip install pytesseract
```

在这里，将使用图 3-1 所示的样品收据图像：

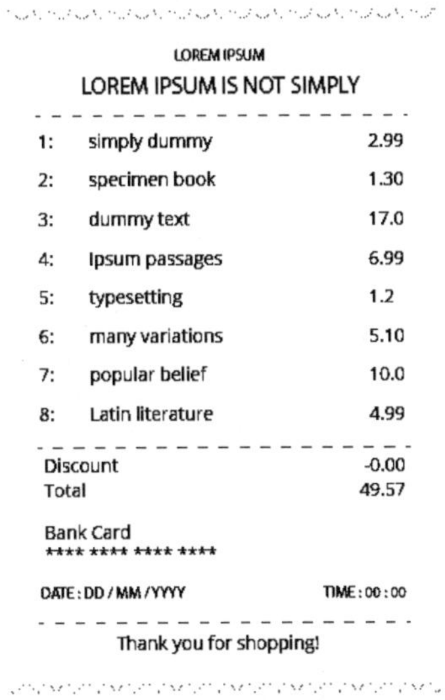

图 3-1 样品收据图像

OCR 第一部分是图像阈值化处理。以下是可用于阈值化的代码：

```python
# importing modules
import cv2

import pytesseract

# reading image using opencv
image = cv2.imread('sample_image.png')

#converting image into gray scale image
gray_image = cv2.cvtColor(image, cv2.COLOR_BGR2GRAY)

# converting it to binary image by Thresholding
# this step is require if you have colored image because if you skip this part
# then tesseract won't able to detect text correctly and this will give incorrect result
threshold_img = cv2.threshold(gray_image, 0, 255, cv2.THRESH_BINARY | cv2.THRESH_OTSU)[1]

# display image
cv2.imshow('threshold image', threshold_img)

# Maintain output window until user presses a key
cv2.waitKey(0)

# Destroying present windows on screen
cv2.destroyAllWindows()
```

图像阈值化处理后，输出图像如图 3-2 所示。

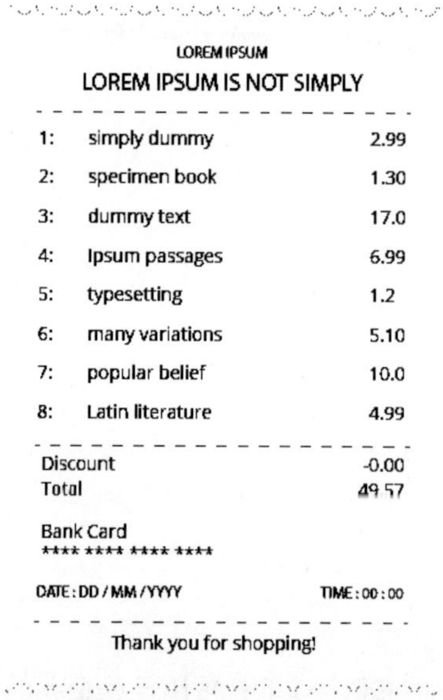

图 3-2　阈值化处理后的图像

现在,可以看到原始图像和阈值图像之间的差异。阈值图像显示了白色像素和黑色像素之间的清晰分离。因此,如果将此图像提供给 Tesseract,它将轻松检测文本区域并提供更准确的结果。为此,请按照下面给出的命令进行操作:

```
#configuring parameters for tesseract

custom_config = r'--oem 3 --psm 6'

# now feeding image to tesseract

details = pytesseract.image_to_data(threshold_img, output_type=Output.DICT, config=custom_config, lang='eng')

print(details.keys())
```

如果要打印详细信息,下列是包含相关详细信息的字典键:

```
dict_keys(['level', 'page_num', 'block_num', 'par_num', 'line_num', 'word_num', 'left', 'top', 'width', 'height', 'conf', 'text'])
```

上面的字典有输入图像的信息,例如检测到的文本区域、位置信息、高度、宽度、置信度等。现在,使用上面的字典在原始图像上绘制边界框,以了解 Tesseract 的工作准确度。作为文本扫描仪来检测文本区域,可按照下面给出的代码实现。

```
total_boxes = len(details['text'])

for sequence_number in range(total_boxes):

    if float(details['conf'][sequence_number]) >30.0:

        (x, y, w, h) = (details['left'][sequence_number], details['top'][sequence_number], details['width'][sequence_number], details['height'][sequence_number])

        threshold_img = cv2.rectangle(threshold_img, (x, y), (x + w, y + h), (0, 255, 0), 2)

# display image

cv2.imshow('captured text', threshold_img)

# Maintain output window until user presses a key

cv2.waitKey(0)

# Destroying present windows on screen

cv2.destroyAllWindows()
```

注意:这里只考虑那些置信度分数大于 30 的图像。在此之后,需要验证所有文本结果是否正确,即使它们的置信度分数为 30~40,也仍需要验证这一点,因为图像混合了数字、其他字符和文本,并且没有向 Tesseract 指定一个字段只有文本或只有数字。还应将整个文档原样提供给 Tesseract 并等待它根据值显示结果,无论它是属于文本还是数字。

现在有了带有边界框的图像，就可以继续下一步，将捕获的文本排列到带有格式的文件中，以便轻松跟踪值。

下面给出的代码是根据当前图像将结果文本排列成一种格式：

```
parse_text = []

word_list = []

last_word = ''

for word in details['text']:

    if word!='':

        word_list.append(word)

        last_word = word

    if (last_word!='' and word == '') or (word==details['text'][-1]):

        parse_text.append(word_list)

        word_list = []
```

下面的代码会将结果文本转换为文件：

```
import csv

with open('result_text.txt', 'w', newline="") as file:

    csv.writer(file, delimiter=" ").writerows(parse_text)
```

3.2　Tesseract 的预处理

为了避免 Tesseract 的输出精度下降，需要确保对图像进行适当的预处理。这包括重新缩放、二值化、去除噪声、去歪斜等。

要为 OCR 预处理图像，请使用以下任意一个 Python 函数：

```
import cv2
import numpy as np

img = cv2.imread('image.jpg')

# get grayscale image
def get_grayscale(image):
    return cv2.cvtColor(image, cv2.COLOR_BGR2GRAY)

# noise removal
def remove_noise(image):
```

```python
    return cv2.medianBlur(image,5)

#thresholding
def thresholding(image):
    return cv2.threshold(image, 0, 255, cv2.THRESH_BINARY + cv2.THRESH_OTSU)[1]

#dilation
def dilate(image):
    kernel = np.ones((5,5),np.uint8)
    return cv2.dilate(image, kernel, iterations = 1)

#erosion
def erode(image):
    kernel = np.ones((5,5),np.uint8)
    return cv2.erode(image, kernel, iterations = 1)

#opening - erosion followed by dilation
def opening(image):
    kernel = np.ones((5,5),np.uint8)
    return cv2.morphologyEx(image, cv2.MORPH_OPEN, kernel)

#canny edge detection
def canny(image):
    return cv2.Canny(image, 100, 200)

#skew correction
def deskew(image):
    coords = np.column_stack(np.where(image > 0))
    angle = cv2.minAreaRect(coords)[-1]
     if angle < -45:
        angle = -(90 + angle)
    else:
        angle = -angle
    (h, w) = image.shape[:2]
    center = (w // 2, h // 2)
    M = cv2.getRotationMatrix2D(center, angle, 1.0)
    rotated  =  cv2.warpAffine(image,  M,  (w,  h),  flags=cv2.INTER_CUBIC, borderMode=cv2.BORDER_REPLICATE)
    return rotated

#template matching
def match_template(image, template):
    return cv2.matchTemplate(image, template, cv2.TM_CCOEFF_NORMED)
```

第 4 章

OpenCV 深度学习

本章首先介绍基于 OpenCV 和深度学习的人脸识别，然后介绍对象检测。

4.1 使用 OpenCV DNN 模块进行深度学习

深度神经网络（Deep Neural Network，DNN）模块是 OpenCV 中的模块，负责与深度学习相关的所有事情。

该模块允许使用来自流行框架（如 TensorFlow、PyTorch 等）的预训练神经网络，并可直接在 OpenCV 中使用这些框架。

这意味着可以使用流行的框架（如 TensorFlow）训练模型，然后仅使用 OpenCV 进行推理与预测。

以下是使用 OpenCV 进行推理时可能的一些优势。
- 通过使用 OpenCV 的 DNN 模块进行推理，最终代码更加紧凑和简单。
- 不熟悉训练框架的人也可以使用该模型。
- 在某些情况下，使用 OpenCV 的 DNN 模块将提供更快的 CPU 推理结果。
- 除了支持基于 CUDA 的 NVIDIA 的 GPU，OpenCV 的 DNN 模块还支持基于 OpenCL 的 Intel GPU。
- 最重要的是，摆脱训练框架不仅使代码更简单，而且最终摆脱了整个框架，这意味着不必使用像 TensorFlow 这样的繁重框架来构建最终应用程序。当尝试在资源受限的边缘设备上部署时，这是一个巨大的优势，例如一个树莓派。

可能让人失望的一件事是 OpenCV 不能用于训练深度学习网络。这可能听起来很糟糕，但不必担心，对于训练神经网络，不应该使用 OpenCV，而有其他专门的库，如 TensorFlow、PyTorch 等来完成这项任务。

OpenCV DNN 模块支持许多流行的深度学习框架。以下是 OpenCV DNN 模块支持的深度学习框架。

1）Caffe

要将预训练的 Caffe 模型与 OpenCV DNN 一起使用，需要两个文件：一个是包含预训练权重的 model.caffemodel 文件；另一个是具有 .prototxt 扩展名的模型架构文件。它就像一个纯文本文件，具有类似 JSON 的结构，包含所有神经网络层的定义。

2）TensorFlow

为了加载预训练的 TensorFlow 模型，还需要两个文件：模型权重文件和包含模型配置的 protobuf 文本文件。模型权重文件有一个.pb 扩展名，它是一个包含所有预训练权重的 protobuf 文件。如果之前使用过 TensorFlow，就会知道.pb 文件是在保存模型并冻结权重后得到的模型检查点；模型配置保存在 protobuf 文本文件中，该文件具有.pbtxt 文件扩展名。

3）Torch 和 PyTorch

为了加载 Torch 模型文件，我们需要包含预训练权重的文件。通常，此文件具有.t7 或.net 扩展名。但是对于具有.pth 扩展名的最新 PyTorch 模型，首先转换为 ONNX 是继续进行的最佳方式。转换为 ONNX 后，可以直接加载它们，因为 OpenCV DNN 支持 ONNX 模型。

4）Darknet

OpenCV DNN 模块也支持著名的 Darknet 框架。通常，要加载 Darknet 模型，需要一个具有.weights 扩展名的模型权重文件。对于 Darknet 模型，网络配置文件将始终是有.cfg 扩展名的文件。

通常，使用 OpenCV DNN 模块进行深度学习时需要执行以下 4 个步骤。

（1）阅读图像和目标类。

（2）使用架构和模型参数初始化 OpenCV DNN 模块。

（3）使用模块对图像执行前向传递。

（4）对结果进行后处理。

对于不同的任务，预处理和后处理步骤是不同的。

首先，应导入所需的库。

```
# Importing Required libraries
import numpy as np
import time
import cv2
import matplotlib.pyplot as plt
import os
import sys
```

然后从加载类名开始，将在 ImageNet 中定义的 1000 个类进行分类。所有这些类都在名为 synset_words.txt 的文本文件中。在这个文本文件中，每个类都在一个具有其唯一 ID 的新行中，而且每个类都有多个标签，例如查看文本文件中的前 3 行：

```
'n01440764 tench, Tinca tinca'
'n01443537 goldfish, Carassius auratus'
'n01484850 great white shark, white shark'
```

因此，对于每一行，都有 Class ID，然后有多个类名，它们都是该类的有效名称，我们只使用第一个。因此，为了做到这一点，我们必须从每一行中提取第二个单词并创建一个新列表，这将是我们的标签列表。

```
# Split all the classes by a new line and store it in variable called rows
rows = open('model/synset_words.txt').read().strip().split("n")
```

```
# Check the number of classes
print("Number of Classes "+str(len(rows)))

# Show the first 5 rows
print(rows[0:5])
```

程序输出如下：

```
Number of Classes 1000 ['n01440764 tench, Tinca tinca', 'n01443537 goldfish, Carassius
auratus', 'n01484850 great white shark, white shark, man-eater, man-eating shark,
Carcharodon carcharias', 'n01491361 tiger shark, Galeocerdo cuvieri', 'n01494475
hammerhead, hammerhead shark']
```

在这里，将提取标签（每行的第二个元素）并创建一个标签列表。

```
# Split by comma after first space is found and grabb the first element and store it
# in a new list
CLASSES = [r[r.find(" ") + 1:].split(",")[0] for r in rows]

# Print the first 20 processed class labels
print(CLASSES[0:20])
```

程序输出如下：

```
['tench', 'goldfish', 'great white shark', 'tiger shark', 'hammerhead', 'electric ray',
'stingray', 'cock', 'hen', 'ostrich', 'brambling', 'goldfinch', 'house finch', 'junco',
'indigo bunting', 'robin', 'bulbul', 'jay', 'magpie', 'chickadee']
```

现在，在可以使用 OpenCV DNN 模块之前，必须使用以下函数之一对其进行初始化。
- Caffe 模型：cv2.dnn.readNetFromCaffe。
- Tensorflow 模型：cv2.dnn.readNetFromTensorFlow。
- Pytorch 模型：cv2.dnn.readNetFromTorch。

由此可见，所使用的函数取决于模型训练的原始架构。

由于我们将使用 Caffe 训练的 DenseNet121，因此我们所用的函数将是：

```
retval = cv2.dnn.readNetFromCaffe( prototxt[, caffeModel] )
```

参数：
- prototxt：.prototxt 文件的路径，这是模型架构的文本描述。
- caffeModel：.caffemodel 文件的路径，这是实际训练的神经网络模型，它包含模型的所有权重/参数。这通常有几兆字节大小。

初始化 OpenCV DNN 模块的代码如下：

```
# Load the Model Weights
weights = 'model/DenseNet_121.caffemodel'

# Load the Model Architecture
architecture ='model/DenseNet_121.prototxt.txt'

# Here we are reading pre-trained caffe model with its architecture
net = cv2.dnn.readNetFromCaffe(architecture,weights)
```

下面读入一个示例图像并使用 matplotlib 模块的 imshow()函数显示它。

```
# Load the input image
```

```
image = cv2.imread('images/jemma2.jpg')

# Display the image
plt.figure(figsize=(10,10))
plt.imshow(image[:,:,::-1]);plt.axis("off");
```

现在，在网络中传递图像之前，还需要对其进行预处理，这意味着将图像调整为训练时的大小，对于许多网络，这是 224×224，在预处理步骤中，还可以执行其他操作，例如归一化图像（使强度值范围为 0～1）和平均减法等。

幸运的是，在 OpenCV 中，有一个 cv2.dnn.blobFromImage()函数，大部分时间它会处理所有的预处理。

```
blob = cv2.dnn.blobFromImage(image[, scalefactor[, size[, mean[, swapRB[, crop]]]]])
```

参数：
- image：输入图像。
- scalefactor：用于归一化图像。该值乘以图像，值为 1 表示不进行缩放。
- size：图像将被调整到的大小，这取决于每个模型。
- mean：这些是整个数据集的平均 R、G、B 通道值，它们分别被从图像的 R、G、B 通道值中减去，这使模型具有光照不变性。
- swapRB：布尔标志（默认为 false），这表示是否交换 3 通道图像中的第一个和最后一个通道是必要的。
- crop：指示调整大小后是否裁剪图像的标志。

```
blob = cv2.dnn.blobFromImage(image, 0.017, (224, 224), (103.94,116.78,123.68))
```

在经过这个函数的处理后，我们得到一个 4 维二进制类型的对象，这就是我们将传递给网络的内容。

```
print(blob.shape)
```

程序输出如下：

```
(1, 3, 224, 224)
```

在这里，将 blob 图像设置为网络的输入。

```
# Passing the blob as input through the network
net.setInput(blob)
```

实际计算将在此处进行。在这里，图像将通过所有模型参数，最后，将获得分类器的输出。

```
Output = net.forward()
# Length of the number of predictions
print("Total Number of Predictions are: {}".format(len(Output[0])))
```

程序输出：

```
Total Number of Predictions are: 1000
# Look at few outputs
Output[0][:4]
```

程序输出：

```
array([[[-2.0572357 ]], [[-0.18754716]], [[-3.314731 ]], [[-6.196114 ]]], dtype=float32)
```

通过查看输出，可以看出模型为每个类别返回了一组分数，但我们需要的是每个类别的值

0~1 的概率。可以通过对分数应用 softmax()函数来获得它们。实现代码如下：

```
# Reshape the Output so its a single dimensional vector
new_Output = Output.reshape(len(Output[0][:]))

# Convert the scores to class probabilities between 0-1 by applying softmax
expanded = np.exp(new_Output - np.max(new_Output))
prob =  expanded / expanded.sum()
# Look at few probablities
prob[:4]
```

程序输出：

```
array([5.7877337e-06, 3.7540856e-05, 1.6458317e-06, 9.2260699e-08], dtype=float32)
```

最大概率是目标类的置信度。

```
# Maximum Probabilitry for this image
conf = np.max(prob)
print(conf)
```

程序输出：

```
0.59984004
```

包含最大置信度/概率的索引是目标类的索引。

```
# Index of Class with the maximum Probability
index = np.argmax(prob)
print(index)
```

程序输出：

```
331
```

通过将上面的索引放入标签列表中，可以获得目标类的名称。

```
# Name of the Class with the maximum probability
label = CLASSES[index]
print(label)
```

程序输出：

```
hare
```

我们已经成功得到了分类，现在可使用所拥有的信息注释图像。实现代码如下：

```
imagec = image.copy()

cv2.putText(imagec, "Label: {}, {:.2f}%".format(label, conf *100), (5, 50), cv2.FONT_HERSHEY_COMPLEX, 2, (100, 20, 255), 2)

plt.figure(figsize=(10,10))
plt.imshow(imagec[:,:,::-1]);plt.axis("off");
```

现在已经逐步讲解了如何使用 OpenCV DNN 模块创建分类管道，下面将创建一个函数来一步完成上述的所有工作。简而言之，将创建以下两个函数。

- 初始化函数：此函数包含将设置一次的网络部分，如加载模型。
- 主要函数：此函数包含从预处理到后处理的所有其余代码，还可以选择返回图像或使用 matplotlib 显示它。

初始化函数将运行一次，它将使用所需文件初始化网络。实现代码如下：

```
def init_classify(weights_name = 'DenseNet_121.caffemodel', architecture_name =
'DenseNet_121.prototxt.txt'):

    # Set global variables
    global net, classes

    base_path = 'model'

    # Read the Classes
    rows = open(os.path.join(base_path,'synset_words.txt')).read().strip().split("n")

    # Load and split the classes
    classes = [r[r.find(" ") + 1:].split(",")[0] for r in rows]

    # Load the wieght and architeture of the model
    weights = os.path.join(base_path, weights_name)
    architecture = os.path.join(base_path, architecture_name)

    # Intialize the model
    net = cv2.dnn.readNetFromCaffe(architecture, weights)
```

主要函数实现代码如下：

```
def classify(image, returndata=False, size=1,):

    # Pre-process the image
    blob = cv2.dnn.blobFromImage(image, 0.017, (224, 224), (103.94,116.78,123.68))

    # Input blob image into network
    net.setInput(blob)

    # Forward pass
    Output = net.forward()

    # Reshape the Output so its a single dimensional vector
    new_Output = Output.reshape(len(Output[0][:]))

    # Convert the scores to class probabilities between 0-1
    expanded = np.exp(new_Output - np.max(new_Output))
    prob = expanded / expanded.sum()

    # Get Highest probable class
    conf= np.max(prob)

    # Index of Class with the maximum Probability
    index = np.argmax(prob)

    # Name of the Class with the maximum probability
```

```python
        label = classes[index]

        text = "Label: {}, {:.2f}%".format(label, conf*100)

        cv2.putText(image, text, (5, size*26), cv2.FONT_HERSHEY_COMPLEX, size, (100, 20, 255), 3)

        if returndata:
            return image, text

        else:
            plt.figure(figsize=(10,10))
            plt.imshow(image[:,:,::-1]);plt.axis("off");
```

当需要对视频进行分类时，将 returndata 设置为 True。

初始化分类器：

```
init_classify()
```

现在可以调用分类器函数并在多个图像上进行测试。

```
image = cv2.imread('images/spcar.JPG')
classify(image)
image = cv2.imread('images/cat.jpeg')
classify(image,size=2)
```

如果想实时使用这个分类器，那么代码如下。

```python
# Initialize fps to 0
fps = 0

# Load the classes and model
init_classify()

# Initialize the webcame
cap=cv2.VideoCapture(0)

while(True):

    # Fetch the inital time in order to find FPS
    start_time = time.time()

    # Read the frame from camera
    ret, frame=cap.read()

    # If camera is not working break the loop
    if not ret:
        break

    # Flip the frame, laterally
    image = cv2.flip(frame,1)

    # Classify the objects in frames
```

```
    image,_ = classify(image, returndata=True)

    # Display the classified object of frames
    cv2.putText(image, 'FPS:{:.2f}'.format(fps), (470, 20),cv2.FONT_HERSHEY_COMPLEX,
0.8, (255, 20, 55), 2)

    # Display the window
    cv2.imshow("Image",image)

    # Calculate FPS
    fps = (1.0 / (time.time() - start_time))

    # Press `q` in order to close the window
    k = cv2.waitKey(1)
    if k == ord('q'):
        break

# Release the camera and destroy all opened windows
cap.release()
cv2.destroyAllWindows()
```

4.2 基于 OpenCV 和深度学习的人脸识别

人脸识别是一种使用人脸识别或验证个人身份的方法。有多种算法可以进行人脸识别，它们的准确性也有所不同。本节描述如何使用深度学习进行人脸识别。

如何使用深度学习识别人脸？可以利用人脸嵌入，其中每张人脸都被转换为一个向量，这种技术称为深度度量学习。可以进一步把这个过程分成 3 个简单的步骤，以便于理解。

1）人脸检测

人脸检测指检测图像或视频流中的人脸，即通过人的确切位置/坐标，提取这张人脸以进行进一步处理。

2）特征提取

人脸识别中的特征提取，即从图像中裁剪的人脸中提取特征。在这里，我们将从嵌入的人脸中提取人脸特征。神经网络将人脸图像作为输入，并输出代表人脸最重要特征的向量。在机器学习中，这个向量被称为嵌入，因此在人脸识别中称这个向量为人脸嵌入。那么，如何识别不同人的面孔呢？

在训练神经网络时，网络学习为看起来相似的人脸输出相似的向量。例如，如果一个人在不同的时间跨度内有多个人脸图像，当然，脸部的某些特征可能会发生变化，但不会发生太大变化。所以在这种情况下，与人脸相关的向量是相似的，或者说，它们在向量空间中非常接近。

现在在训练网络后，网络学习输出同一个人（看起来相似）的面孔的向量彼此更接近（相似）。

在这里不会训练这样的网络，因为训练这样的网络需要大量的数据和计算能力。本节将使用由 Davis King 在约 300 万张图像数据集上训练的预训练网络。该网络输出一个由 128 个数字组成的向量，这些数字代表了人脸最重要的特征。

现在已经知道了这个网络是如何工作的，下面来看看如何在我们自己的数据中使用这个网络：将数据中的所有图像传递给这个预训练的网络以获取相应的嵌入，并将这些嵌入保存在文件中以备下一步使用。

3）比较人脸

现在已经在文件中保存了所嵌入的每张人脸的数据，下一步是识别不在数据中的新图像。因此，首先是使用和在上面使用的相同网络计算图像的人脸嵌入，然后将此嵌入与已经拥有的其余嵌入进行比较。如果生成的嵌入与任何其他嵌入更接近或相似，就会识别出人脸。

接下来，将使用 OpenCV 和 Python 实现人脸识别。可以通过 Anaconda 在 Windows 上安装 dlib。

```
>conda install -c conda-forge dlib
```

接下来安装 face_recognition：

```
>pip install face_recognition
```

现在已经安装了所有依赖项，可以开始编码了。首先创建 3 个文件。一个将使用数据集并使用 dlib 为每张人脸提取人脸嵌入。接下来，将这些嵌入保存在一个文件中。

在下一个文件中，将人脸与图像中现有的识别人脸进行比较，做同样的事情，在实时网络摄像头源中识别人脸。

首先，需要获得一个数据集，或者创建一个属于自己的数据集。只需确保将所有图像排列在文件夹中，每个文件夹仅包含一个人的图像。

从数据集提取人脸特征的代码如下：

```python
from imutils import paths
import face_recognition
import pickle
import cv2
import os

#get paths of each file in folder named Images
#Images here contains my data(folders of various persons)
imagePaths = list(paths.list_images('Images'))
knownEncodings = []
knownNames = []
# loop over the image paths
for (i, imagePath) in enumerate(imagePaths):
    # extract the person name from the image path
    name = imagePath.split(os.path.sep)[-2]
    # load the input image and convert it from BGR (OpenCV ordering)
    # to dlib ordering (RGB)
    image = cv2.imread(imagePath)
    rgb = cv2.cvtColor(image, cv2.COLOR_BGR2RGB)
    #Use Face_recognition to locate faces
    boxes = face_recognition.face_locations(rgb,model='hog')
    # compute the facial embedding for the face
    encodings = face_recognition.face_encodings(rgb, boxes)
    # loop over the encodings
    for encoding in encodings:
        knownEncodings.append(encoding)
        knownNames.append(name)
#save emcodings along with their names in dictionary data
```

```
data = {"encodings": knownEncodings, "names": knownNames}
#use pickle to save data into a file for later use
f = open("face_enc", "wb")
f.write(pickle.dumps(data))
f.close()
```

现在已经将嵌入存储在一个名为"face_enc"的文件中，可以使用它们来识别图像或实时视频流中的人脸。

用于检测和识别图像中人脸的代码如下：

```
import face_recognition
import imutils
import pickle
import time
import cv2
import os

#find path of xml file containing haarcascade file
cascPathface = os.path.dirname(
 cv2.__file__) + "/data/haarcascade_frontalface_alt2.xml"
# load the harcaascade in the cascade classifier
faceCascade = cv2.CascadeClassifier(cascPathface)
# load the known faces and embeddings saved in last file
data = pickle.loads(open('face_enc', "rb").read())
#Find path to the image you want to detect face and pass it here
image = cv2.imread(Path-to-img)
rgb = cv2.cvtColor(image, cv2.COLOR_BGR2RGB)
#convert image to Greyscale for haarcascade
gray = cv2.cvtColor(image, cv2.COLOR_BGR2GRAY)
faces = faceCascade.detectMultiScale(gray,
                                     scaleFactor=1.1,
                                     minNeighbors=5,
                                     minSize=(60, 60),
                                     flags=cv2.CASCADE_SCALE_IMAGE)

# the facial embeddings for face in input
encodings = face_recognition.face_encodings(rgb)
names = []
# loop over the facial embeddings incase
# we have multiple embeddings for multiple fcaes
for encoding in encodings:
    #Compare encodings with encodings in data["encodings"]
    #Matches contain array with boolean values and True for the embeddings it matches
    #closely and False for rest
    matches = face_recognition.compare_faces(data["encodings"],
    encoding)
    #set name =inknown if no encoding matches
    name = "Unknown"
    # check to see if we have found a match
    if True in matches:
        #Find positions at which we get True and store them
```

```python
        matchedIdxs = [i for (i, b) in enumerate(matches) if b]
        counts = {}
        # loop over the matched indexes and maintain a count for
        # each recognized face face
        for i in matchedIdxs:
            #Check the names at respective indexes we stored in matchedIdxs
            name = data["names"][i]
            #increase count for the name we got
            counts[name] = counts.get(name, 0) + 1
            #set name which has highest count
            name = max(counts, key=counts.get)


    # update the list of names
    names.append(name)
    # loop over the recognized faces
    for ((x, y, w, h), name) in zip(faces, names):
        # rescale the face coordinates
        # draw the predicted face name on the image
        cv2.rectangle(image, (x, y), (x + w, y + h), (0, 255, 0), 2)
        cv2.putText(image, name, (x, y), cv2.FONT_HERSHEY_SIMPLEX,
            0.75, (0, 255, 0), 2)
    cv2.imshow("Frame", image)
cv2.waitKey(0)
```

4.3 英特尔 OpenVINO 工具包简介

神经网络架构的训练是大多数参与深度学习领域的人的动力。我们无休止地关注数据的数量、质量以及应该使用什么样的神经网络架构。但这真的是正确的方法吗？优化模型以进行部署同样重要，因为它将加快推理速度。英特尔 OpenVINO 工具包可让用户做到这一点。当目标设备是英特尔硬件（如英特尔 CPU 和 GPU）或边缘设备（如 OAK-D）时，它是神经网络模型优化的最佳工具之一。

因此，本节将全面讲解英特尔 OpenVINO 工具包——从安装到运行预训练模型，在英特尔 CPU 上近乎实时地使用 OpenVINO 进行优化。

什么是 OpenVINO？OpenVINO 代表开放视觉推理和神经网络优化，它的功能正如其名称所暗示的那样。它优化了我们训练有素的神经网络模型，以便我们可以非常有效地使用它们进行推理。

英特尔 OpenVINO 工具包是一个开源工具，不仅有助于优化深度学习模型，还可以使用这些模型开发解决方案，并确保它们的快速部署。这还不是全部。再加上优化的图像和视频处理管道，OpenVINO 工具包还可以继续在边缘提供快速推理。

虽然，它不支持神经网络的训练，但是，我们已经有很多好的深度神经网络训练框架。此外，重要的是要强调 OpenVINO 工具包所做的一切，即大规模优化和深度学习推理，它做得最好。

现在开始熟悉英特尔 OpenVINO 工具包并专注基于视觉的任务。

英特尔 OpenVINO 优化模型显然在英特尔公司的硬件上运行得最好，包括：
- 广泛的英特尔至强可扩展和英特尔酷睿处理器，也包括配备英特尔高清显卡的英特尔奔腾处理器，甚至英特尔凌动处理器。
- 英特尔 iGPU（集成 GPU），如英特尔 Iris、英特尔高清显卡和英特尔 UHD 卡。
- FPGA，如 Intel Arria 10 FPGA GX。
- 由 Myriad X VPU 提供支持的英特尔神经计算棒 2 代。

除了英特尔神经计算棒 2 代之外，OAK-1 和 OAK-D 是由 Myriad X VPU 提供支持的另外两个设备，它们使用了 OpenVINO 的优化模型。

OAK-1 和 OAK-D 都使用 OpenVINO 的中间表示（Intermediate Representation，IR）文件来获取它们支持的最终模型文件（.blob 文件）。

一个工具包服务这么多设备！难怪英特尔 OpenVINO 工具包如今被认为是提供深度神经网络解决方案的最佳选择之一。它不仅可以以非常轻松地将被部署的计算机视觉应用程序扩展到多个摄像头，而且可以通过访问官方链接 https://www.intel.com/content/www/us/en/developer/tools/openvino-toolkit/system-requirements.html 访问每个工具包要求和支持的硬件。

FP32 和 INT8 模型，无论是来自英特尔官方模型动物园还是公共存储库，都在 CPU 上运行得最好。尽管 CPU 也支持 FP16 模型，但 FP16 模型在 GPU 上的性能最好。

安装英特尔 OpenVINO 工具包的步骤如下。

1）下载正确版本的 OpenVINO 工具包

用户可根据需要和操作系统选择所有先决条件后，前往官方下载页面 https://www.intel.com/content/www/us/en/developer/tools/openvino-toolkit/download.html 获取正确版本的 OpenVINO。

应该下载哪个版本？任何较旧的可用版本都可以使用，但最好使用最新版本，因为它提供了工具包的最新功能。

2）解压安装程序文件

打开终端并输入 cd 命令，进入保存 OpenVINO 安装程序文件的目录。然后解压文件，使用以下命令：

```
tar -xvzf l_openvino_toolkit_p_<version>.tgz
```

解压后，将在当前工作目录中看到一个 l_openvino_toolkit_p_<version>文件夹。这包含 OpenVINO 安装程序和其他所需文件。

3）安装 OpenVINO 工具包

首先，输入 cd 命令，进入 l_openvino_toolkit_p_<version>，然后可以通过三个选项在系统上安装 OpenVINO。
- 图形用户界面 (GUI) 安装向导。
- 命令行安装程序。
- 命令行安装程序，带有无提示说明。

我们将使用 GUI 安装向导，即 install_GUI.sh 文件，因为这种方式最简单和最直观。

开始安装，在终端中键入以下命令：

```
sudo ./install_GUI.sh
```

4）按照屏幕上的安装说明进行操作

首先看到的是欢迎屏幕，然后是许可协议。接下来，将看到先决条件屏幕。

即使收到不满足先决条件的警告，仍然可以继续安装。这些可以在以后安装依赖项时安装。

之后是安装配置摘要屏幕。如果愿意，还可以自定义配置，并选择要安装的内容。

如果以管理员权限按照上面列出的步骤安装 OpenVINO 工具包，它将被安装在 /opt/intel/openvino_<version>/目录中。

接下来的几个步骤可确保在使用此工具包时不会遇到任何问题。无论是否要使用 OpenVINO 的任何高级功能，这些步骤都是必要的。

5）安装软件依赖项

其中包括以下软件依赖项：
- 英特尔优化的 OpenCV 库构建。
- 深度学习推理引擎。
- 深度学习模型优化工具。

将当前工作目录更改为 install_dependencies 文件夹。

```
cd /opt/intel/openvino_2021/install_dependencies
```

运行以下脚本来安装依赖项。

```
sudo -E ./install_openvino_dependencies.sh
```

6）设置环境变量

要设置环境变量，需要打开一个新终端并输入：

```
vi ~/.bashrc
```

转到文件末尾，并添加以下行：

```
source /opt/intel/openvino_2021/bin/setupvars.sh
```

现在保存并关闭文件，然后关闭当前终端并打开一个新终端，以便进行系统范围的更改。

7）安装模型优化器先决条件

模型优化器是 OpenVINO 工具包中的一个命令行工具。使用此模型优化器将使用不同框架训练的模型转换为 OpenVINO 工具包所接受的用于推理的格式。

供您参考，OpenVINO 工具包不直接支持推理。使用 TensorFlow、Caffe、MXNet、ONNX 甚至 Kaldi 等深度学习框架训练的模型都无法进行推理，而首先需要将这些经过训练的模型转换为 IR，其中包括：
- 一个描述网络架构的.xml 文件。
- 一个.bin 文件，其中包含训练模型的权重和偏差。

为了进行转换，需要在安装必要的先决条件后通过模型优化器运行模型。

为了配置模型优化器，请打开终端并转到模型优化器先决条件目录。

```
cd /opt/intel/openvino_2021/deployment_tools/model_optimizer/install_prerequisites
```

执行以下命令运行脚本，一次性为 Caffe、TensorFlow、MXNet、ONNX 甚至 Kaldi 配置模型优化器。

```
sudo ./install_prerequisites.sh
```

这样就完成了 OpenVINO 工具包的安装过程。现在，已准备好使用英特尔模型动物园中的任何模型或转换模型以进行推理。

现在来关注模型转换过程。在这里，将讲解如何将预训练的深度学习模型从支持框架转换为 OpenVINO IR 格式。这些是 OpenVINO 工具包所支持的框架：Caffe、TensorFlow、MXNet、Kaldi、PyTorch 和 ONNX。

具体来说，将涵盖以下模型的转换：
- Caffe 转换为 OpenVINO IR 格式。
- TensorFlow 转换为 OpenVINO IR 格式。

下面从一个图像分类模型 SqueezeNet Caffe 开始介绍转换过程。SqueezeNet Caffe 模型是英特尔公司模型动物园中的公开可用模型之一。

按照以下简单步骤开始将 SqueezeNet Caffe 模型转换为 OpenVINO IR 格式：

（1）从英特尔公司公共模型动物园中下载 SqueezeNet Caffe 模型。
（2）运行模型优化器将 Caffe 模型转换为 IR 格式。

首先下载 SqueezeNet Caffe 模型。该模型在 open_model_zoo 目录下的 public 子目录中命名为 squeezenet1.1。只需提供模型名称，同时执行 downloader.py 脚本以确保下载正确的模型。

现在转到 deployment_tools 中的 tools/downloader 目录。

```
cd /opt/intel/openvino_2021/deployment_tools/open_model_zoo/tools/downloader
```

在这里，使用以下命令执行 downloader.py 脚本：

```
python3 downloader.py --name squeezenet1.1
```

执行后，将在当前工作目录中看到一个 public 文件夹，其中包含 squeezenet1.1 子目录。它由 squeezenet1.1.caffemodel、squeezenet1.1.prototxt 和 squeezenet1.1.prototxt.orig. 3 个文件组成：

其中，我们最感兴趣的是 squeezenet1.1.caffemodel 和 squeezenet1.1.prototxt 文件。下面需要它们来运行模型优化器并获取.xml 和.bin 文件。
- .caffemodel 文件包含模型权重。
- .prototxt 文件包含模型优化器所需的模型架构。

接下来，运行模型优化器并将 Caffe 模型转换为 IR 格式。因此，转到 model_optimizer 目录。

```
cd /opt/intel/openvino_2021_3_latest/openvino_2021/deployment_tools/model_optimizer
```

接下来，执行以下命令：

```
python mo.py --input_model squeezenet1.1.caffemodel --batch 1 --output_dir squeezenet_ir
```

让我们回顾一下使用的标志：
- --input_model：这是要转换为 IR 格式的 Caffe 模型的路径。在上面的示例中，已假设 Caffe 模型与 mo.py 脚本位于同一目录中。请注意，即使是 squeezenet1.1.prototxt 文件也应该存在于同一目录中，以便脚本可以自行推断文件的路径；否则，必须使用 --input_proto 标志提供.prototxt 文件的路径。
- --batch：此标志指定构建 OpenVINO 模型的批量大小。它在推理过程中发挥作用。默认情况下，批量大小为 1。批量大小决定了模型在执行推理脚本时将推断出的图像或帧的数量。
- --output_dir：这是存储生成的.xml 和.bin 文件的输出目录。如果不存在，系统将自动创建该目录。

如果一切运行成功，应该会看到类似以下的输出：

```
Note that install_prerequisites scripts may install additional components.
[ SUCCESS ] Generated IR version 10 model.
[ SUCCESS ] XML file: /home/sovit/my_data/Data_Science/Projects/openvino_experiments/squeezenet1.1_caffemodel/squeezenet_ir/squeezenet1.1.xml
[ SUCCESS ] BIN file: /home/sovit/my_data/Data_Science/Projects/openvino_experiments/
```

```
squeezenet1.1_caffemodel/squeezenet_ir/squeezenet1.1.bin
    [ SUCCESS ] Total execution time: 6.39 seconds.
    [ SUCCESS ] Memory consumed: 344 MB.
```

在 squeezenet_ir 目录中，会找到需要的两个文件。

（1）squeezenet1.1.bin：该文件包含模型权重。

（2）squeezenet1.1.xml：该文件包含模型优化器所需的拓扑/架构。

我们已经成功地将第一个图像分类 Caffe 模型转换为适当的 IR 格式，现在可以利用它在英特尔公司的硬件上运行推理。

接下来，将 TensorFlow 对象检测模型转换为 OpenVINO IR 格式。为此，将使用来自公共模型的 YOLOV3 Tiny TF 模型，遵循与 Caffe 模型转换相同的步骤，会在运行模型优化器时处理一些额外的配置文件和标志。

首先下载模型。

```
python3 downloader.py --name yolo-v3-tiny-tf
```

下载完成后，会在 downloader/public 目录中找到一个 yolo-v3-tiny-tf 文件夹。它包含

（1）yolo-v3-tiny-tf.json 文件，即模型配置文件。

（2）yolo-v3-tiny-tf.pb 文件，这个文件包含模型权重。

接下来，执行模型优化器来获取 .xml 和 .bin 文件。

```
python mo.py --input_model yolo-v3-tiny-tf.pb --transformations_config yolo-v3-tiny-tf.json --batch 1 --output_dir yolo_v3_tiny_ir
```

在这里，可以看到一个额外的 --transformations_config 标志。通过这个标志标示接收 .json 文件的路径，就像在上面的代码块中提供的那样。生成的 OpenVINO 模型权重和拓扑文件将保存在 yolo_v3_tiny_ir 文件夹中。在此目录中，可以找到 yolo-v3-tiny-tf.bin 和 yolo-v3-tiny-tf.xml 两个文件。

下面来将 Tiny YOLOv4 权重文件转换为 OpenVINO-IR 格式。当从原始 FP32 模型转换为 OpenVINO 优化的 FP32 模型时，请观察神经网络层如何变化。

首先从官方 Darknet 存储库 https://github.com/AlexeyAB/darknet 下载 yolov4-tiny.weights，即原始的 Tiny YOLOv4 Darknet 权重文件。

不能通过模型优化器直接使用 Darknet 权重文件来获取 OpenVINO IR 文件，而应首先将权重文件转换为 TensorFlow .pb 文件。该文件是一个冻结模型，其中包含图定义和权重。转换为 .pb 文件有如下好处。

- 除了权重，该文件还有所有的变量操作，转换为存储权重值的常量。
- 冻结模型还有助于将其部署在 Web 服务器和边缘设备上，因为它现在更容易优化。层的融合就是一个这样的例子——许多操作可以一起计算。
- 此外，.pb 格式仅包含保存的模型权重，丢弃所有梯度、元数据和训练变量。这会减小最终文件的大小，从而更容易导出和提供模型。

按照存储库 https://github.com/TNTWEN/OpenVINO-YOLOV4#yolov4-tiny 中给出的步骤，现在将 Darknet 权重转换为 TensorFlow .pb 文件。步骤非常简单，运行转换命令后，应该就会获得一个 frozen_darknet_yolov4_model.pb 文件。

最后，使用模型优化器获得优化后的 FP32 模型，该模型为所需的 OpenVINO IR 格式。在运行命令之前，请确保在同一目录中具有：freeze_darknet_yolov4_model.pb 文件和存储库提供

的 yolo_v4_tiny.json 文件。

接下来，只需给出这个命令，就像在转换 Tiny YOLOv3 模型时所做的那样：

```
python mo.py --input_model frozen_darknet_yolov4_model.pb --transformations_config yolo_v4_tiny.json --batch 1
```

转换完成后，应该在同一目录下获得 frozen_darknet_yolov4_model.bin 和 frozen_darknet_yolov4_model.xml 文件。

OpenVINO 工具包在将原始 FP32 模型转换为 IR 文件时进行了一些优化。虽然这些优化有助于模型对图像和视频流进行更快的推理，但它往往会在一定程度上降低模型的准确性。

那么，现在可以使用 COCO mAP 评估器检查 Tiny YOLOv4 Darknet 权重和 OpenVINO 优化的 FP32 模型的准确性。

为 Tiny YOLOv4 模型计算 mAP 的先决条件有：

- 安装用于 mAP 评估的 COCO API。为此，需要复制存储库 https://github.com/cocodataset/cocoapi。进入 coco/PythonAPI 目录，打开终端并运行 make。
- 需要 MS COCO 2017 验证集（图像和注解），因此请从其官方网站 https://cocodataset.org/ 下载数据集。

在验证图像并使用 COCO API 运行评估后，使用 Tiny YOLOv4 Darknet 权重，得到以下结果：

```
Average Precision  (AP) @[ IoU=0.50:0.95 | area=   all | maxDets=100 ] = 0.185
Average Precision  (AP) @[ IoU=0.50      | area=   all | maxDets=100 ] = 0.334
Average Precision  (AP) @[ IoU=0.75      | area=   all | maxDets=100 ] = 0.185
```

对于 IoU=0.50:0.95，平均精度为 0.185。

现在观察一下 OpenVINO 优化如何影响 Tiny YOLOv4 模型。为了比较，以下是在 Tiny YOLOv4 FP32 模型上运行评估时的结果。

```
Average Precision  (AP) @[ IoU=0.50:0.95 | area=   all | maxDets=100 ] = 0.152
Average Precision  (AP) @[ IoU=0.50      | area=   all | maxDets=100 ] = 0.275
Average Precision  (AP) @[ IoU=0.75      | area=   all | maxDets=100 ] = 0.151
```

可以发现，OpenVINO 优化模型在 IoU=0.50:0.95 的情况下平均精度较低，即平均精度为 0.152。

4.4 使用深度学习的自动车牌识别

深度学习已经成为人们日常生活的一部分，从语音助手到自动驾驶汽车，它无处不在。一种这样的应用是自动车牌识别（Automatic License Plate Recognition，ALPR）。顾名思义，ALPR 是一种利用人工智能和深度学习的理论自动检测和识别车辆牌照字符的技术。

本节将重点介绍 ALPR 的端到端实现。它将侧重于车牌检测和检测到的车牌的 OCR 两步过程。

想象一个美丽的夏天，你在高速公路上开车，收音机里播放着你最喜欢的歌曲，你越过限速，在 70km/h 的限速区以 90km/h 的速度驶过几个摄像头，然后意识到你的错误，但为时已晚。几周后，你会收到一张罚单，上面附有你的汽车图像的证据。你一定想知道：交管部门是否手动检查每张图片并发送罚单？

当然不是，那是 ALPR 系统发送的。从捕获的图像或镜头中，ALPR 检测并提取你的车牌号并向你发送罚单。这一切都是因为简单的 ALPR 系统和几行代码。

ALPR 或 ANPR（Automatic Number Plate Recognition）是负责使用光学字符识别在图像或视频序列中读取车辆牌照的技术。随着深度学习和计算机视觉技术的最新进展，这些任务可以在几毫秒内完成。

ALPR 是一个非常先进和智能的系统，因其具有各种现实生活用例和无须任何人为干预而广受欢迎。让我们来看看它们吧！

（1）交通违规：警察或执法部门可以使用 ALPR 来识别车辆在交通信号灯处仅使用一个摄像头的任何交通违规行为。它还可用于实时识别任何被盗或未注册的车辆。

（2）停车管理：很多地方的停车管理需要大量的人机交互，但是由于有了 ALPR，人机交互可以减少到零。安装在停车场的摄像头可以识别入口处的车牌并将其存储在数据库中，或者仅当车辆在数据库中时才允许车辆进入。退出时，可以重新识别车牌并进行相应的收费。

（3）收费站付款：在高速公路上，人工收费站会变得很忙，并可能导致巨大的交通流量。使用 ANPR，收费站可以识别车牌并自动接收付款。

ALPR 是被广泛使用的计算机视觉应用之一。它利用了各种方法，如对象检测、OCR、分割等。对于硬件，ALPR 系统只需要一个摄像头和一个好的 GPU。为简单起见，这里将重点介绍其检测和识别两步的实现过程。

（1）检测：首先，将视频序列的图像或帧从摄像头或已存储的文件传递给检测算法，该算法检测车牌并返回该车牌的边界框位置。

（2）识别：将 OCR 应用于检测到的车牌，识别车牌的字符，并以文本格式以相同的顺序返回字符。输出可以存储在数据库中，也可以绘制在图像上以进行可视化。

下面让我们详细地看一下每个步骤。

4.4.1 使用 YOLOv4 检测车牌

该管道模块负责从视频序列的图像或帧中检测车牌。

检测过程可以使用任何检测器完成，无论是基于区域的检测器还是单次检测器。本节将重点介绍一种称为 YOLOv4 的单次检测器，主要是因为它具有良好的速度和精度折中以及更好的检测小物体的能力。YOLOv4 将使用 Darknet 框架实现。

Darknet 是一个用 C 语言和统一计算设备架构（Compute Unified Device Architecture，CUDA）编写的开源神经网络框架。YOLOv4 使用 CSPDarknet53 CNN，这意味着它的目标检测主干使用了共有 53 个卷积层的 Darknet53。Darknet 非常易于安装、使用，只需几行代码即可完成。

```
!git clone https://github.com/AlexeyAB/darknet
```

安装和编译 Darknet，并根据环境需要设置一些参数。实现代码如下：

```
%cd darknet
!sed -i 's/OPENCV=0/OPENCV=1/' Makefile
!sed -i 's/GPU=0/GPU=1/' Makefile
!sed -i 's/CUDNN=0/CUDNN=1/' Makefile
!sed -i 's/CUDNN_HALF=0/CUDNN_HALF=1/' Makefile
!sed -i 's/LIBSO=0/LIBSO=1/' Makefile
```

```
!make
```

在这里，一些参数（如 OpenCV、GPU、CUD NN 等）设置为1，即设置为 True，因为它们是提高代码效率和更快地运行计算所必需的。

为了训练 YOLOv4 检测器，将使用 Google 的车辆开放图像数据集。谷歌的"开放图像"是一个开源数据集，包含数千张带有注释的对象图像，用于对象检测、分割等。

让我们看一下数据集。

```
import math
# Creating a list of image files of the dataset
data_path = './data/obj/train/'
files = os.listdir(data_path)
img_arr = []

# Displaying 4 images only
num = 4

# Appending the array of images to a list
for fimg in files:
    if fimg.endswith('.jpg'):
        demo = img.imread(data_path+fimg)
        img_arr.append(demo)
        if len(img_arr) == num:
            break

# Plotting the images using matplotlib
_, axs = plt.subplots(math.floor(num/2), math.ceil(num/2), figsize=(50, 28))

axs = axs.flatten()

for cent, ax in zip(img_arr, axs):
    ax.imshow(cent)
plt.show()
```

为了让模型学习，需要在数据集上训练它。在开始训练过程之前，需要修改配置文件（.cfg）。需要修改的参数是批量大小、细分、类等。

创建两个文件：一个是 obj.data，其中包含训练数据、测试数据和类信息的信息；另一个是 obj.names，其中包含所有类的名称。

下一步是下载 YOLOv4 的预训练权重。

```
!wget https://github.com/AlexeyAB/darknet/releases/download/darknet_yolo_v3_optimal/yolov4.conv.137
```

现在是训练的重要部分。

```
!./darknet detector train data/obj.data cfg/yolov4-obj.cfg yolov4.conv.137 -dont_show -map
```

参数包括 obj.data 文件、配置文件和 yolov4 预训练权重。

现在车牌检测器已经完全训练好了，是时候使用它了。为此，将创建一个名为 yolo_det() 的辅助函数。该函数负责从输入的车辆图像中检测车牌的边界框。实现代码如下：

```
def yolo_det(frame, config_file, data_file, batch_size, weights, threshold, output,
```

```python
network, class_names, class_colors, save = False, out_path = ''):

    prev_time = time.time()

    # Preprocessing the input image
    width = darknet.network_width(network)
    height = darknet.network_height(network)
    darknet_image = darknet.make_image(width, height, 3)
    image_rgb = cv2.cvtColor(frame, cv2.COLOR_BGR2RGB)
    image_resized = cv2.resize(image_rgb, (width, height))

    # Passing the image to the detector and store the detections
    darknet.copy_image_from_bytes(darknet_image, image_resized.tobytes())
    detections = darknet.detect_image(network, class_names, darknet_image, thresh=threshold)
    darknet.free_image(darknet_image)

    # Plotting the deetections using darknet in-built functions
    image = darknet.draw_boxes(detections, image_resized, class_colors)
    print(detections)
    if save:
        im = cv2.cvtColor(image, cv2.COLOR_RGB2BGR)
        file_name = out_path + '-det.jpg'
        cv2.imwrite(os.path.join(output, file_name), im)

    # Calculating time taken and FPS for detection
    det_time = time.time() - prev_time
    fps = int(1/(time.time() - prev_time))
    print("Detection time: {}".format(det_time))

    # Resizing predicted bounding box from 416x416 to input image resolution
    out_size = frame.shape[:2]
    in_size = image_resized.shape[:2]
    coord, scores = resize_bbox(detections, out_size, in_size)
    return coord, scores, det_time
```

让我们再定义一个函数 resize_bbox()，用于根据原始图像大小将预测的边界框坐标调整回边界框坐标。当图像通过检测器时，应将图像大小调整到一定的分辨率。检测器还以相同的分辨率输出结果，在本例中为 416×416 像素。要将图像转换回其输入分辨率，需要相应地更改边界框坐标。实现代码如下：

```python
def resize_bbox(detections, out_size, in_size):
    coord = []
    scores = []

    # Scaling the bounding boxes according to the original image resolution
    for det in detections:
        points = list(det[2])
        conf = det[1]
```

```
        xmin, ymin, xmax, ymax = darknet.bbox2points(points)
        y_scale = float(out_size[0]) / in_size[0]
        x_scale = float(out_size[1]) / in_size[1]
        ymin = int(y_scale * ymin)
        ymax = int(y_scale * ymax)
        xmin = int(x_scale * xmin) if int(x_scale * xmin) > 0 else 0
        xmax = int(x_scale * xmax)

        final_points = [xmin, ymin, xmax-xmin, ymax-ymin]
        scores.append(conf)
        coord.append(final_points)

    return coord, scores
```

4.4.2 OCR

训练自定义车牌检测器后,而进入 ALPR 的文本识别。文本识别是通过理解和分析其潜在模式从场景中识别文本的过程。它也被称为 OCR。它还可以用于各种应用,如文档阅读、信息检索、货架产品识别等。OCR 可以被训练或用作预训练模型。本节将使用一个预训练的 OCR 模型。

PaddleOCR 就是这样一种用于 OCR 的框架或工具包。PaddleOCR 为用户提供多语言实用 OCR 工具,帮助用户在几行代码中应用和训练不同的模型。PaddleOCR 在其工具包中提供了很多模型,包括 PP-OCR、一系列高质量的预训练 OCR、最新的算法如 SRN,以及流行的 OCR 算法如 CRNN。

PaddleOCR 还提供了不同类型的模型,包括轻量级(占用较少内存)的模型、重量级(占用大量内存)的模型,以及可自由使用的预训练权重。

使用这些模型,可在几行代码中实现 PaddleOCR,并将为 ALPR 系统创造奇迹。

首先,安装所需的工具包和依赖项。这些依赖项和工具有助于访问 OCR 实施所需的所有文件和脚本。

```
# Navigating to previous directory or home directory
%cd ../
# Installing dependencies
!pip install paddlepaddle-gpu
!pip install "paddleocr>=2.0.1"
```

安装完成后,OCR 需要根据用户的要求进行初始化。

```
from paddleocr import PaddleOCR
ocr = PaddleOCR(lang='en',rec_algorithm='CRNN')
```

使用 PaddleOCR 初始化 OCR,需要以下几个参数:
- Lang:指定要识别的语言。
- det_algorithm:指定使用的文本检测算法。
- Rec_algorithm:指定使用的识别算法。

对于 ALPR,只会传递两个参数,即语言和识别算法。在这里,使用了 lang 作为英语和 CRNN 识别算法,CRNN 识别算法在这个工具包中也称为 PPOCRv2。

只需一行代码即可使用此 OCR。

```
# Syntax
result = ocr.ocr(cr_img, cls=False, det=False)
```

这里，cr_img 是传递给 OCR 的图像，cls 和 det 是设置为 False 的参数，因为在 ALPR 管道中不需要文本检测器和文本角度分类器。

现在车牌检测器已经完全训练好了，OCR 已经准备就绪，是时候将所有这些放在一起并投入使用了。为此，将创建一些辅助函数来一次性访问所有功能。

首先，将创建一个函数 crop()，该函数通过将图像和坐标作为参数来负责裁剪图像。

```
def crop(image, coord):
    # Cropping is done by -> image[y1:y2, x1:x2]
    cr_img = image[coord[1]:coord[3], coord[0]:coord[2]]
    return cr_img
```

为了对图像执行 ANPR，我们将创建一个最终函数，比如 test_img()，它将在一个地方执行检测、裁剪、OCR 和输出绘图。

在此之前，将初始化一些变量：

```
# Variables storing colors and fonts
font = cv2.FONT_HERSHEY_SIMPLEX
blue_color = (255,0,0)
white_color = (255,255,255)
black_color = (0,0,0)
green_color = (0,255,0)
yellow_color = (178, 247, 218)

def test_img(input, config_file, weights, out_path):
    # Loading darknet network and classes along with the bbox colors
    network, class_names, class_colors = darknet.load_network(
            config_file,
            data_file,
            weights,
            batch_size= batch_size
        )

    # Reading the image and performing YOLOv4 detection.
    img = cv2.imread(input)
    bboxes, scores, det_time = yolo_det(img, config_file, data_file, batch_size, weights, thresh, out_path, network, class_names, class_colors)

    # Extracting or cropping the license plate and applying the OCR
    for bbox in bboxes:
        bbox = [bbox[0], bbox[1], bbox[2]- bbox[0], bbox[3] - bbox[1]]
        cr_img = crop(img, bbox)
        result = ocr.ocr(cr_img, cls=False, det=False)
        ocr_res = result[0][0]
        rec_conf = result[0][1]
```

```
    # Plotting the predictions using OpenCV
    (label_width,label_height), baseline = cv2.getTextSize(ocr_res , font, 2, 3)
    top_left = tuple(map(int,[int(bbox[0]),int(bbox[1])-(label_height+baseline)]))
    top_right = tuple(map(int,[int(bbox[0])+label_width,int(bbox[1])]))
    org = tuple(map(int,[int(bbox[0]),int(bbox[1])-baseline]))

    cv2.rectangle(img, (int(bbox[0]), int(bbox[1])), (int(bbox[2]), int(bbox[3])),
blue_color, 2)
    cv2.rectangle(img, top_left, top_right, blue_color,-1)
    cv2.putText(img, ocr_res, org, font, 2, white_color,3)

    # Writing output image
    file_name = os.path.join(out_path, 'out_' + input.split('/')[-1])
    cv2.imwrite(file_name, img)
```

4.5 超分辨率

超分辨率（Super Resolution，SR）是指放大或改善图像细节的过程。当增加图像的尺寸时，需要以某种方式对额外的像素进行插值。基本的图像处理技术不会产生良好的结果，因为它们在放大时不会将周围环境纳入上下文。深度学习和生成对抗网络（Generative Adversarial Networks，GAN）在这里提供了帮助，并提供了更好的结果。

OpenCV 目前提供 4 种深度学习算法供选择用于放大图像。下面将讨论的 4 种方法是 EDSR、ESPCN、FSRCNN、LapSRN。请注意，前 3 种算法提供了 2、3 和 4 倍的放大率，而最后一种算法提供原始大小的 2、4 和 8 倍的放大率。

为了将上面列出的模型用于超分辨率，需要使用标准 OpenCV 模块的附加功能。这就是为什么必须安装 opencv-contrib 模块的原因。此外，超分辨率存在于模块 dnn_superres 中。该模块在 C++的 OpenCV 4.1 版和 Python 的 OpenCV 4.3 版中实现。

如果已经安装了 OpenCV，则可以使用以下代码片段检查其版本：

```
import cv2
print(cv2.__version__)
```

如果 OpenCV 版本早于 4.3，则可以使用以下命令对其进行升级：

```
pip install opencv-contrib-python --upgrade
```

如果没有安装 OpenCV，则可以通过命令直接使用 pip 安装最新版本：

```
pip install opencv-contrib-python
```

为了比较上述算法，下面使用图 4-1 作为参考进行具体说明，来尝试生成图像右上角的 OpenCV 徽标的高分辨率图像，以了解超分辨率 OpenCV super-res 模块的功能。

```
import cv2
import matplotlib.pyplot as plt
# Read image
img = cv2.imread("AI-Courses-By-OpenCV-Github.png")
plt.imshow(img[:,:,::-1])
plt.show()
```

图 4-1

首先导入 OpenCV 和 Matplotlib 模块并读取测试图像。使用下面给出的代码裁剪出 OpenCV 徽标。

```
# Cropout OpenCV logo
img = img[:80,850:]
plt.imshow(img[:,:,::-1])
plt.show()
```

使用 EDSR 算法实现超分辨率的代码如下：

```
sr = cv2.dnn_superres.DnnSuperResImpl_create()

path = "EDSR_x4.pb"

sr.readModel(path)

sr.setModel("edsr",4)

result = sr.upsample(img)

# Resized image
resized = cv2.resize(img,dsize=None,fx=4,fy=4)

plt.figure(figsize=(12,8))
plt.subplot(1,3,1)
# Original image
plt.imshow(img[:,:,::-1])
plt.subplot(1,3,2)
# SR upscaled
plt.imshow(result[:,:,::-1])
plt.subplot(1,3,3)
# OpenCV upscaled
plt.imshow(resized[:,:,::-1])
plt.show()
```

ESPCN 的基本结构受 SRCNN 的启发，使用亚像素卷积层代替常规卷积层，其作用类似于反卷积层。最后一层使用亚像素卷积层来生成高分辨率地图。与此同时，Shi 等发现 Tanh 激活函数比标准 ReLu 函数效果更好。

使用 ESPCN 算法实现超分辨率的代码如下：

```
sr = cv2.dnn_superres.DnnSuperResImpl_create()

path = "ESPCN_x3.pb"
```

```
sr.readModel(path)

sr.setModel("espcn",3)

result = sr.upsample(img)

# Resized image
resized = cv2.resize(img,dsize=None,fx=3,fy=3)

plt.figure(figsize=(12,8))
plt.subplot(1,3,1)
# Original image
plt.imshow(img[:,:,::-1])
plt.subplot(1,3,2)
# SR upscaled
plt.imshow(result[:,:,::-1])
plt.subplot(1,3,3)
# OpenCV upscaled
plt.imshow(resized[:,:,::-1])
plt.show()
```

FSRCNN 和 ESPCN 的概念非常相似。它们都具有受 SRCNN 启发的基本结构,并在最后使用上采样层来提高速度,而不是在早期对其进行插值。此外,它们甚至在最终使用更多映射层之前缩小输入特征维度并使用更小的滤波器尺寸,这导致模型更小也更快。

该架构从卷积层开始,其滤波器大小从 SRCNN 的 9 元素降至 5 元素。应用收缩层是因为输入分辨率本身可能很大并且需要很长时间。使用 1×1 的滤波器大小,不会增加计算成本。

使用 FSRCNN 算法实现超分辨率的代码如下:

```
sr = cv2.dnn_superres.DnnSuperResImpl_create()

path = "FSRCNN_x3.pb"

sr.readModel(path)

sr.setModel("fsrcnn",3)

result = sr.upsample(img)

# Resized image
resized = cv2.resize(img,dsize=None,fx=3,fy=3)

plt.figure(figsize=(12,8))
plt.subplot(1,3,1)
# Original image
plt.imshow(img[:,:,::-1])
plt.subplot(1,3,2)
# SR upscaled
plt.imshow(result[:,:,::-1])
plt.subplot(1,3,3)
```

```
# OpenCV upscaled
plt.imshow(resized[:,:,::-1])
plt.show()
```

LapSRN 模型由两个分支组成：特征提取和图像重建分支。参数共享发生在不同尺度之间，即 4x 使用来自 2x 模型的参数等。这意味着一个金字塔用于缩放 2x，两个用于 4x，三个用于 8x。制作如此深的模型意味着它们可能会遇到梯度消失问题。因此，尝试了不同类型的本地跳过连接，例如不同源跳过连接和共享源连接。模型的损失函数使用 Charbonnier()，没有使用批归一化（Batch Narmalizution）。

使用 LapSRN 算法实现超分辨率的代码如下：

```
sr = cv2.dnn_superres.DnnSuperResImpl_create()

path = "LapSRN_x8.pb"

sr.readModel(path)

sr.setModel("lapsrn",8)

result = sr.upsample(img)

# Resized image
resized = cv2.resize(img,dsize=None,fx=8,fy=8)

plt.figure(figsize=(12,8))
plt.subplot(1,3,1)
# Original image
plt.imshow(img[:,:,::-1])
plt.subplot(1,3,2)
# SR upscaled
plt.imshow(result[:,:,::-1])
plt.subplot(1,3,3)
# OpenCV upscaled
plt.imshow(resized[:,:,::-1])
plt.show()
```

超分辨率的应用遍布各个领域。

- 医学成像：超分辨率是提高 X 射线、CT 扫描等质量的绝佳解决方案。它有助于突出有关人体解剖和功能信息的重要细节。提高分辨率或增强医学图像也有助于突出关键堵塞或肿瘤。
- 生物特征识别：超分辨率可以通过增强人脸、指纹和虹膜图像在生物特征识别中发挥关键作用。形状、结构和纹理大大增强，有助于清晰识别生物特征痕迹。
- 遥感：在遥感和卫星成像中使用超分辨率技术已经发展了几十年。事实上，第一个超分辨率想法的动机是需要更高质量和分辨率的 Landsat 遥感图像。
- 天文成像：提高天文图片的分辨率有助于关注微小的细节，这些细节可能会成为外太空的重大发现。
- 监控成像：交通监控和安全系统在维护平民安全方面发挥着非常重要的作用。对数字录制的视频应用超分辨率在识别交通或安全违规方面大有帮助。

4.6 对象检测

对象检测是计算机视觉和图像处理中的一项任务，用于检测图像或视频中的对象。它用于各种现实世界的应用，包括视频监控、自动驾驶汽车、物体跟踪等。

例如，要使汽车真正实现自动驾驶，它必须识别并跟踪周围的物体（如汽车、行人和交通灯），主要的信息来源之一是使用物体检测的摄像头。最重要的是，检测应该是实时的，这需要相对较快的方式，以便汽车可以安全地在街道上行驶。

本节将讲解如何使用 YOLOv3 算法和 OpenCV 或 Python 中的 PyTorch 实现对象检测。

在深入研究代码之前，先安装所需的库：

```
>pip3 install opencv-python numpy matplotlib
```

如果想要使用 PyTorch 代码，那么安装 PyTorch 可以参考 https://pytorch.org/get-started/locally/。

从头开始构建 YOLOv3 整个系统（模型和使用的技术）非常具有挑战性，Darknet 或 OpenCV 等开源库已经为你构建好了。

导入所需模块：

```
import cv2
import numpy as np

import time
import sys
import os
```

定义一些需要的变量和参数：

```
CONFIDENCE = 0.5
SCORE_THRESHOLD = 0.5
IOU_THRESHOLD = 0.5

# the neural network configuration
config_path = "cfg/yolov3.cfg"
# the YOLO net weights file
weights_path = "weights/yolov3.weights"
# weights_path = "weights/yolov3-tiny.weights"

# loading all the class labels (objects)
labels = open("data/coco.names").read().strip().split("\n")
# generating colors for each object for later plotting
colors = np.random.randint(0, 255, size=(len(LABELS), 3), dtype="uint8")
```

上面的代码初始化了参数，config_path 和 weights_path 分别代表模型配置（即 yolov3）和对应的预训练模型权重。标签是要检测的不同对象的所有类标签的列表，下面将用唯一的颜色绘制每个对象类，这就是生成随机颜色的原因。

加载模型的代码如下：

```
# load the YOLO network
net = cv2.dnn.readNetFromDarknet(config_path, weights_path)
```

加载一个示例图像的代码如下：

```python
path_name = "images/street.jpg"
image = cv2.imread(path_name)
file_name = os.path.basename(path_name)
filename, ext = file_name.split(".")
```

接下来,需要对该图像进行归一化、缩放和重塑,使其适合作为神经网络的输入:

```python
h, w = image.shape[:2]
# create 4D blob
blob = cv2.dnn.blobFromImage(image, 1/255.0, (416, 416), swapRB=True, crop=False)
```

这会将像素值归一化为 0~1 的范围,将图像大小调整为(416,416),并对其进行重塑:

```python
print("image.shape:", image.shape)
print("blob.shape:", blob.shape)
```

输出:

```
image.shape: (1200, 1800, 3)
blob.shape: (1, 3, 416, 416)
```

现在可将此图像输入神经网络以获得输出预测:

```python
# sets the blob as the input of the network
net.setInput(blob)
# get all the layer names
ln = net.getLayerNames()
ln = [ln[i[0] - 1] for i in net.getUnconnectedOutLayers()]
# feed forward (inference) and get the network output
# measure how much it took in seconds
start = time.perf_counter()
layer_outputs = net.forward(ln)
time_took = time.perf_counter() - start
print(f"Time took: {time_took:.2f}s")
```

这将提取神经网络输出并打印推理所需的总时间:

```
Time took: 1.54s
```

现在你可能想知道,为什么它没有那么快?1.5s 很慢?好吧,我们仅将 CPU 用于推理,这对于实际问题并不理想,这就是将在后面介绍 PyTorch 的原因。另一方面,与 R-CNN 等其他技术相比,1.5s 相对较好。

现在需要迭代神经网络输出并丢弃任何置信度低于之前指定的 CONFIDENCE 参数(即 0.5 或 50%)的对象。

```python
font_scale = 1
thickness = 1
boxes, confidences, class_ids = [], [], []
# loop over each of the layer outputs
for output in layer_outputs:
    # loop over each of the object detections
    for detection in output:
        # extract the class id (label) and confidence (as a probability) of
        # the current object detection
        scores = detection[5:]
        class_id = np.argmax(scores)
```

```python
        confidence = scores[class_id]
        # discard out weak predictions by ensuring the detected
        # probability is greater than the minimum probability
        if confidence > CONFIDENCE:
            # scale the bounding box coordinates back relative to the
            # size of the image, keeping in mind that YOLO actually
            # returns the center (x, y)-coordinates of the bounding
            # box followed by the boxes' width and height
            box = detection[:4] * np.array([w, h, w, h])
            (centerX, centerY, width, height) = box.astype("int")
            # use the center (x, y)-coordinates to derive the top and
            # and left corner of the bounding box
            x = int(centerX - (width / 2))
            y = int(centerY - (height / 2))
            # update our list of bounding box coordinates, confidences,
            # and class IDs
            boxes.append([x, y, int(width), int(height)])
            confidences.append(float(confidence))
            class_ids.append(class_id)
```

这将遍历所有预测并仅保存具有高置信度的对象,让我们看看检测向量代表什么:

```python
print(detection.shape)
```

输出:

```
(85,)
```

在每个对象预测上,有一个长度为 85 的向量。前 4 个值表示对象的位置,(x, y) 坐标为中心点以及边界框的宽度和高度,其余数字对应于对象标签,因为这是 COCO 数据集,它有 80 个类标签。

例如,如果检测到的对象是人,则长度为 80 的向量中的第一个值应为 1,其余所有值应为 0,对于自行车则是第二个数字,对于汽车则是第三个数字,一直到第 80 个对象。这就是使用 np.argmax() 函数来获取类 ID 的原因,因为这个函数返回长度为 80 的向量中最大值的索引。

现在已经有了所需要的一切,下面来绘制对象矩形和标签并查看结果:

```python
# loop over the indexes we are keeping
for i in range(len(boxes)):
    # extract the bounding box coordinates
    x, y = boxes[i][0], boxes[i][1]
    w, h = boxes[i][2], boxes[i][3]
    # draw a bounding box rectangle and label on the image
    color = [int(c) for c in colors[class_ids[i]]]
    cv2.rectangle(image, (x, y), (x + w, y + h), color=color, thickness=thickness)
    text = f"{labels[class_ids[i]]}: {confidences[i]:.2f}"
    # calculate text width & height to draw the transparent boxes as background of the
    # text
    (text_width, text_height) = cv2.getTextSize(text, cv2.FONT_HERSHEY_SIMPLEX,
fontScale=font_scale, thickness=thickness)[0]
    text_offset_x = x
    text_offset_y = y - 5
    box_coords = ((text_offset_x, text_offset_y), (text_offset_x + text_width + 2,
```

```
text_offset_y - text_height))
    overlay = image.copy()
    cv2.rectangle(overlay, box_coords[0], box_coords[1], color=color, thickness=
cv2.FILLED)
    # add opacity (transparency to the box)
    image = cv2.addWeighted(overlay, 0.6, image, 0.4, 0)
    # now put the text (label: confidence %)
    cv2.putText(image, text, (x, y - 5), cv2.FONT_HERSHEY_SIMPLEX,
        fontScale=font_scale, color=(0, 0, 0), thickness=thickness)
```

保存图像：

```
cv2.imwrite(filename + "_yolo3." + ext, image)
```

图像中存在单个对象的两个边界框，可以使用一种称为非最大值抑制的技术来消除这种情况。

非最大值抑制是一种抑制重叠边界框的技术。主要分两个阶段实现非最大值抑制：
- 选择获得最高置信度（即概率）的边界框。
- 然后将所有其他边界框与这个选定的边界框进行比较，并消除那些具有高 IoU 的边界框。

重叠度（Intersection over Union，IoU）是一种在非最大值抑制中使用的技术，用于比较两个不同边界框的接近程度。图 4-2 简单演示了 IoU 计算方法。

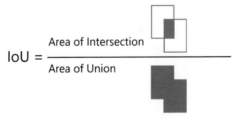

图 4-2　IoU 计算方法

IoU 越高，边界框越近。IoU 为 1 意味着两个边界框是相同的，而 IoU 为 0 意味着它们没有相交。

因此，将使用 0.5 的 IoU 阈值，这意味着与最大概率边界框相比，已消除了低于此值的任何边界框。

SCORE_THRESHOLD 将消除任何置信度低于该值的边界框：

```
# perform the non maximum suppression given the scores defined before
idxs = cv2.dnn.NMSBoxes(boxes, confidences, SCORE_THRESHOLD, IOU_THRESHOLD)
```

再次绘制边界框：

```
# ensure at least one detection exists
if len(idxs) > 0:
    # loop over the indexes we are keeping
    for i in idxs.flatten():
        # extract the bounding box coordinates
        x, y = boxes[i][0], boxes[i][1]
        w, h = boxes[i][2], boxes[i][3]
        # draw a bounding box rectangle and label on the image
        color = [int(c) for c in colors[class_ids[i]]]
        cv2.rectangle(image, (x, y), (x + w, y + h), color=color, thickness=thickness
```

```
        text = f"{labels[class_ids[i]]}: {confidences[i]:.2f}"
        # calculate text width & height to draw the transparent boxes as background
        # of the text
        (text_width, text_height) = cv2.getTextSize(text, cv2.FONT_HERSHEY_SIMPLEX,
fontScale=font_scale, thickness=thickness)[0]
        text_offset_x = x
        text_offset_y = y - 5
        box_coords = ((text_offset_x, text_offset_y), (text_offset_x + text_width +
2, text_offset_y - text_height))
        overlay = image.copy()
        cv2.rectangle(overlay, box_coords[0], box_coords[1], color=color, thickness=
cv2.FILLED)
        # add opacity (transparency to the box)
        image = cv2.addWeighted(overlay, 0.6, image, 0.4, 0)
        # now put the text (label: confidence %)
        cv2.putText(image, text, (x, y - 5), cv2.FONT_HERSHEY_SIMPLEX,
            fontScale=font_scale, color=(0, 0, 0), thickness=thickness)
```

将图像保存到磁盘：

```
cv2.imwrite(filename + "_yolo3." + ext, image)
```

如果想使用 GPU 进行推理，那么可以使用支持 CUDA 计算的 PyTorch 库，代码如下：

```
import cv2
import matplotlib.pyplot as plt
from utils import *
from darknet import Darknet

# Set the NMS Threshold
nms_threshold = 0.6
# Set the IoU threshold
iou_threshold = 0.4
cfg_file = "cfg/yolov3.cfg"
weight_file = "weights/yolov3.weights"
namesfile = "data/coco.names"
m = Darknet(cfg_file)
m.load_weights(weight_file)
class_names = load_class_names(namesfile)
# m.print_network()
original_image = cv2.imread("images/city_scene.jpg")
original_image = cv2.cvtColor(original_image, cv2.COLOR_BGR2RGB)
img = cv2.resize(original_image, (m.width, m.height))
# detect the objects
boxes = detect_objects(m, img, iou_threshold, nms_threshold)
# plot the image with the bounding boxes and corresponding object class labels
plot_boxes(original_image, boxes, class_names, plot_labels=True)
```

4.7 GOTURN：基于深度学习的对象跟踪

本节将讲解一种名为 GOTURN 的基于深度学习的对象跟踪算法。GOTURN 的原始实现是在 Caffe 中，但它已被移植到 OpenCV 跟踪 API，下面将使用这个 API 在 Python 中演示 GOTURN。

对象跟踪的目标是跟踪视频序列中的对象。使用视频序列的帧和边界框来初始化跟踪算法，以指示用户有兴趣跟踪的对象的位置。跟踪算法为所有后续帧输出一个边界框。

GOTURN 是 Generic Object Tracking Using Regression Networks 的缩写，是一种基于深度学习的跟踪算法。

大多数跟踪算法都是在线训练的。换句话说，跟踪算法在运行时学习它正在跟踪的对象的外观。

因此，许多实时跟踪器依赖于通常比基于深度学习的解决方案快得多的在线学习算法。

GOTURN 通过离线学习对象的运动改变了我们将深度学习应用于跟踪问题的方式。GOTURN 模型在数千个视频序列上进行训练，不需要在运行时执行任何学习。

GOTURN 是由 David Held、Sebastian Thrun、Silvio Savarese 在他们题为"Learning to Track at 100 FPS with Deep Regression Networks"的论文中提出的。

GOTURN 使用一对来自数千个视频的裁剪帧进行训练。

在第一帧（也称为前一帧）中，对象的位置是已知的，帧被裁剪为对象周围边界框大小的 2 倍。第一个裁剪帧中的对象始终居中。

需要预测对象在第二帧（也称为当前帧）中的位置。用于裁剪第一帧的边界框也用于裁剪第二帧。因为对象可能已经移动，所以对象不在第二帧的中心。

训练卷积神经网络（CNN）以预测第二帧中边界框的位置。

为了在 OpenCV 中使用 GOTURN，可以下载位于 https://github.com/spmallick/goturn-files 的 GOTURN caffemodel 和 prototxt 文件。

通过 GitHub 共享的 GOTURN 模型文件由于其很大，因此被拆分为 4 个不同的文件。这些文件需要在解压前合并。在 Windows 系统上，可以使用 7-zip 合并文件。

现在来看看跟踪器是如何使用的。

创建跟踪器：首先需要创建一个 GOTURN 跟踪器类的实例。

```
# Create tracker
tracker = cv2.TrackerGOTURN_create()
```

接下来，读取一个视频帧：

```
# Read video
video = cv2.VideoCapture("chaplin.mp4")
# Exit if video not opened
if not video.isOpened():
    print("Could not open video")
    sys.exit()
# Read first frame
ok,frame = video.read()
if not ok:
    print("Cannot read video file")
    sys.exit()
```

定义边界框：需要在视频中选择要跟踪的对象。这是通过定义边界框或选择 ROI 来完成的。在我们的示例中，我们对边界框进行了硬编码，但也可以使用 cv2.selectROI() 调用 GUI 来查找感兴趣的区域。

```
# Define a bounding box
bbox = (276, 23, 86, 320)
```

```
# Uncomment the line below to select a different bounding box
#bbox = cv2.selectROI(frame, False)
```

初始化跟踪器：跟踪器将第一帧和要跟踪的对象周围的边界框作为输入。

```
# Initialize tracker with first frame and bounding box
ok = tracker.init(frame,bbox)
```

预测新帧中的边界框：遍历视频中的所有帧，并使用 tracker.update() 找到新帧的边界框。其余代码仅用于计时和显示。

```
while True:
    # Read a new frame
    ok, frame = video.read()
    if not ok:
        break
    
    # Start timer
    timer = cv2.getTickCount()
    
    # Update tracker
    ok, bbox = tracker.update(frame)
    
    # Calculate Frames per second (FPS)
    fps = cv2.getTickFrequency() / (cv2.getTickCount() - timer);
    
    # Draw bounding box
    if ok:
        # Tracking success
        p1 = (int(bbox[0]), int(bbox[1]))
        p2 = (int(bbox[0] + bbox[2]), int(bbox[1] + bbox[3]))
        cv2.rectangle(frame, p1, p2, (255,0,0), 2, 1)
    else :
        # Tracking failure
        cv2.putText(frame, "Tracking failure detected", (100,80), cv2.FONT_HERSHEY_SIMPLEX, 0.75,(0,0,255),2)
    
    # Display tracker type on frame
    cv2.putText(frame, "GOTURN Tracker", (100,20), cv2.FONT_HERSHEY_SIMPLEX, 0.75, (50,170,50),2);
    
    # Display FPS on frame
    cv2.putText(frame, "FPS : " + str(int(fps)), (100,50), cv2.FONT_HERSHEY_SIMPLEX, 0.75, (50,170,50), 2);
    
    # Display result
    cv2.imshow("Tracking", frame)
    
    # Exit if ESC pressed
    k = cv2.waitKey(1) & 0xff
    if k == 27:
        break
```

与其他基于深度学习的跟踪器相比，GOTURN 速度很快。它在 Caffe 的 GPU 上以 100 帧/秒的速度运行，在 OpenCV CPU 中以大约 20 帧/秒的速度运行。尽管跟踪器是通用的，但理论上可以通过使用特定类型的对象偏置训练集来在特定对象（如行人）上获得更好的结果。

4.8 手势识别

手势识别是人机交互技术中一个活跃的研究领域。它在虚拟环境控制和手语翻译、机器人控制或音乐创作方面有许多应用。在这个关于手势识别的机器学习项目中，我们将使用 OpenCV、Python 中的 MediaPipe 框架和 TensorFlow 制作一个实时手势识别器。

MediaPipe 是由 Google 公司开发的可定制的机器学习解决方案框架。它是一个开源和跨平台的框架，而且非常轻量级。MediaPipe 自带一些预训练的机器学习解决方案，例如人脸检测、姿势估计、手部识别、物体检测等。

首先使用 MediaPipe 来识别手部和手部关键点。MediaPipe 为每只检测到的手返回总共 21 个关键点。

要构建这个手势识别项目，需要 4 个包。首先导入这些包。

```python
# import necessary packages for hand gesture recognition project using Python OpenCV
import cv2
import numpy as np
import mediapipe as mp
import tensorflow as tf
from tensorflow.keras.models import load_model
```

初始化 MediaPipe：

```python
# initialize mediapipe
mpHands = mp.solutions.hands
hands = mpHands.Hands(max_num_hands=1, min_detection_confidence=0.7)
mpDraw = mp.solutions.drawing_utils
```

mp.solution.hands 模块执行手部识别算法，需要创建对象并将其存储在 mpHands 中。

使用 mpHands.Hands()方法配置模型。第一个参数是 max_num_hands，这意味着模型将在单帧中检测到的最大手数。MediaPipe 可以在单个帧中检测多只手，但在这个项目中一次只能检测一只手。

mp.solutions.drawing_utils 会绘制检测到的关键点，这样就不用手动绘制了。

初始化 TensorFlow：

```python
# Load the gesture recognizer model
model = load_model('mp_hand_gesture')

# Load class names
f = open('gesture.names', 'r')
classNames = f.read().split('\n')
f.close()
print(classNames)
```

使用 load_model()函数加载 TensorFlow 预训练模型。

gesture.names 文件包含手势类的名称。所以首先使用 Python 内置的 open()函数打开文件，然后读取文件。

之后，使用 read()函数读取文件。

输出：

```
['okay', 'peace', 'thumbs up', 'thumbs down', 'call me', 'stop', 'rock', 'live long',
'fist', 'smile']
```

该模型可以识别 10 种不同的手势。

从网络摄像头读取帧：

```
# Initialize the webcam for Hand Gesture Recognition Python project
cap = cv2.VideoCapture(0)

while True:
  # Read each frame from the webcam
  _, frame = cap.read()
  x , y, c = frame.shape

  # Flip the frame vertically
  frame = cv2.flip(frame, 1)
  # Show the final output
  cv2.imshow("Output", frame)
  if cv2.waitKey(1) == ord('q'):
      break

# release the webcam and destroy all active windows
cap.release()
cv2.destroyAllWindows()
```

创建一个 VideoCapture 对象并传递一个参数"0"，它是系统的摄像机 ID。在这种情况下，有 1 个网络摄像头与系统连接。如果有多个网络摄像头，则根据摄像头 ID 更改参数；否则，将其保留为默认值。

cap.read()函数从网络摄像头读取每帧。

cv2.flip()函数翻转帧。

cv2.imshow()函数在新的 OpenCV 窗口上显示帧。

cv2.waitKey()函数使窗口保持打开状态，直到按 Q 键。

检测手部关键点：

```
framergb = cv2.cvtColor(frame, cv2.COLOR_BGR2RGB)
# Get hand landmark prediction
result = hands.process(framergb)

className = ''

# post process the result
if result.multi_hand_landmarks:
    landmarks = []
    for handslms in result.multi_hand_landmarks:
```

```
        for lm in handslms.landmark:
            # print(id, lm)
            lmx = int(lm.x * x)
            lmy = int(lm.y * y)

            landmarks.append([lmx, lmy])

        # Drawing landmarks on frames
        mpDraw.draw_landmarks(frame, handslms, mpHands.HAND_CONNECTIONS)
```

MediaPipe 使用 RGB 格式的图像，但 OpenCV 读取 BGR 格式的图像。因此，使用 cv2.cvtCOLOR()函数将帧转换为 RGB 格式。

process()函数接收一个 RGB 帧并返回一个结果类。

然后使用 result.multi_hand_landmarks 列表检查是否检测到任何手。

之后，遍历每个检测并将坐标存储在一个称为 landmarks 的列表中。

最后，使用 mpDraw.draw_landmarks()函数绘制帧中的所有地标。

识别手势：

```
# Predict gesture in Hand Gesture Recognition project
prediction = model.predict([landmarks])
print(prediction)
classID = np.argmax(prediction)
className = classNames[classID]

# show the prediction on the frame
cv2.putText(frame, className, (10, 50), cv2.FONT_HERSHEY_SIMPLEX,
            1, (0,0,255), 2, cv2.LINE_AA)
```

model.predict()函数接收一个地标列表并返回一个数组，其中包含每个地标的 10 个预测类。输出如下：

```
[[2.0691623e-18 1.9585415e-27 9.9990010e-01 9.7559416e-05
  1.6617223e-06 1.0814080e-18 1.1070732e-27 4.4744065e-16 6.6466129e-07 4.9615162e-21]]
```

np.argmax()返回列表中最大值的索引。

获得索引后，可以简单地从 classNames 列表中获取类名。

然后，使用 cv2.putText()函数将检测到的手势显示到帧中。

4.9 人体姿态估计

利用物体检测技术，可以检测到人类，但不能具体指出是哪个人的活动；而利用人体姿态估计技术，可以检测到人体并可分析出该特定人体的姿态。

人体姿态估计需要检测和定位身体的主要部位/关节，如肩膀、脚踝、膝盖、手腕等。

人体姿态估计表示人体的图形骨架，有助于分析人类的活动。骨架基本上是一组描述人的姿势的坐标。每个关节都是一个单独的坐标，称为关键点或姿势地标。关键点之间的连接称为对。

通过姿势估计，能够跟踪人类在现实世界空间中的运动和活动。这开辟了广泛的应用可能性。它是一种强大的技术，有助于有效地构建复杂的应用程序。

1. 人体姿态估计技术的应用

目前，人体姿态估计技术有以下一些实际应用。

1）人类活动估计

人体姿态估计可用于跟踪人类活动，如步行、跑步、睡眠、饮酒。它提供了一些关于一个人的信息。活动估计可以增强安全和监视系统。

2）运动转移

人体姿态估计最有趣的应用之一是运动转移。我们在电影或游戏中看到，三维图形角色的身体运动就像真实的人类或动物。通过跟踪人体姿势，三维渲染的图形可以通过人体的运动进行动画处理。

3）机器人

为了训练机器人的运动，可以使用人体姿态估计。不是手动对机器人进行编程以遵循特定路径，而是使用人体姿势骨架来训练机器人的关节运动。

4）游戏

在虚拟现实游戏中，三维姿势由一个或多个摄像头估计，游戏角色根据人类的动作移动。

2. 人体姿态估计模型

人体姿态估计模型主要有以下三类。

（1）运动学模型。它是一种基于骨骼的模型，代表人体。

（2）平面模型。平面模型是一种基于轮廓的模型，它使用人体周围的轮廓来表示人体形状。

（3）体积模型。体积模型创建人体的三维网格，表示人体的形状和外观。

3. 人体姿势估计的类别

（1）二维姿态估计。在二维姿态估计中，仅预测图像中每个地标的 x 和 y 坐标。它不提供有关骨架角度、物体/人体实例的旋转或方向的任何信息。

（2）三维姿态估计。三维姿态估计允许预测人的螺旋位置。它为每个地标提供 x、y 和 z 坐标。通过三维姿态估计，可以确定人体骨骼每个关节的角度。

（3）刚体姿态估计。刚体姿态估计也称为六维姿态估计。它提供有关人体姿态以及人体实例的旋转和方向的所有信息。

（4）单一姿态估计。在单一姿态估计模型中，只能在图像中预测一个人的姿态。

（5）多姿态估计。在多姿态估计中，可以在一幅图像中同时预测多个人体姿态。

4. 人体姿态估计的方法

人体姿态估计主要使用深度学习解决方案来预测人体姿态标志。它将图像作为输入，并为每个实例提供姿势界标作为输出。

人体姿态估计的方法有两种。

（1）自下而上方法。在这种方法中，在图像中预测特定关键点的每个实例，然后将一组关键点组合成最终骨架。

（2）自上而下方法。在这种的方法中，首先在给定图像中检测对象——人，然后在该图像的每个裁剪对象实例中预测地标。

5. 构建人体姿态估计器

在本节，使用带有 OpenCV 的 MediaPipe 框架来构建人体姿态估计器。

MediaPipe 是谷歌开发的开源框架。它是一个非常轻量级的多平台机器学习解决方案框架，可以在 CPU 上实时运行。

实现项目的步骤如下：

（1）导入依赖项。代码如下：

```
# Human pose estimator
# import necessary packages

import cv2
import mediapipe as mp
```

只需要项目的两个依赖项：第一个是名为 cv2 的 OpenCV；第二个是 mediapipe。

（2）创建检测器对象。代码如下：

```
# initialize Pose estimator
mp_drawing = mp.solutions.drawing_utils
mp_pose = mp.solutions.pose

pose = mp_pose.Pose(
    min_detection_confidence=0.5,
    min_tracking_confidence=0.5)
```

在上面的代码中，已经定义了 mediapipe 的姿态估计对象。

mp.solutions.drawing_utils 将在帧上绘制检测到的骨骼关节和对。

MediaPipe 在后端使用 TensorFlow lite。使用检测器首先在帧内定位人/兴趣区域（Region of Interest，ROI），然后使用 ROI 裁剪帧作为输入并预测 ROI 内的地标/关键点。MediaPipe 姿态估计器共检测到 33 个关键点。

MediaPipe 姿态估计器是一个单一的 3D 姿态估计器。它检测每个地标的 x、y 和 z 坐标。z 轴基本上是关于地标深度的信息。

（3）从视频文件中进行检测。代码如下：

```
# create capture object
cap = cv2.VideoCapture('video.mp4')

while cap.isOpened():
    # read frame from capture object
    _, frame = cap.read()

    try:
        # convert the frame to RGB format
        RGB = cv2.cvtColor(frame, cv2.COLOR_BGR2RGB)

        # process the RGB frame to get the result
        results = pose.process(RGB)
print(results.pose_landmarks)
```

首先，使用 cv2.VideoCapture('video.mp4') 创建了一个捕获对象。这将允许读取视频文件。cap.isOpened() 检查捕获对象是否打开成功。然后使用 cap.read() 从视频文件中读取视频帧，直到打开捕获对象。MediaPipe 适用于 RGB 格式的图像，但是 OpenCV 以 BGR 格式读取图像，因此使用 cv2.cvtColor() 函数将帧转换为 RGB 格式。然后，使用 pose.process() 函数从帧中进行检测。它提供了一个包含有关帧的所有信息的结果类。可以使用 results. pose_landmarks 提取所有关键点坐标。

（4）在视频帧上绘制检测。代码如下：

```
# draw detected skeleton on the frame
    mp_drawing.draw_landmarks(
      frame, results.pose_landmarks, mp_pose.POSE_CONNECTIONS)

    # show the final output
    cv2.imshow('Output', frame)
except:
    break
if cv2.waitKey(1) == ord('q'):
    break
cap.release()
cv2.destroyAllWindows()
```

mp_drawing.draw_landmarks()将直接从结果中绘制所有检测到的对。最后，使用 cv2.imshow() 函数显示最终输出，并将循环内容保存在 try-except 块中。因为，如果捕获对象已经从视频文件中读取了所有帧，然后尝试读取下一帧，那么它将找不到任何要读取的帧，并且会抛出错误。因此，如果在运行时发生任何错误，循环将中断并且程序将关闭而不会引发任何错误。cap.release()函数用于释放捕获对象，而 cv2.destroyAllWindows()用于关闭所有活动的 OpenCV 窗口。

4.10 使用OpenPose在OpenCV中进行多人姿态估计

在本节，将讨论如何执行多人姿态估计。

当一张照片中有多个人时，姿态估计会产生多个独立的关键点，需要弄清楚哪一组关键点属于同一个人。

我们将使用在 COCO 数据集上训练的 18 点模型。COCO 数据集使用的关键点及其编号如下：

鼻子—0，颈部—1，右肩—2，右肘—3，右手腕—4，

左肩—5，左肘—6，左手腕—7，右臀部—8，

右膝—9，右脚踝—10，左臀部—11，左膝—12，

踝关节—13，右眼—14，左眼—15，右耳—16，

左耳—17，背景—18

多人姿态估计模型将大小为 $h×w$ 的彩色图像作为输入，并生成矩阵数组作为输出，该矩阵由关键点的置信度图和每个关键点对的部分亲和力热图组成。置信度图用于查找关键点，亲和图用于获取关键点之间的有效连接。上述网络架构由如下两个阶段组成。

（1）阶段 0：VGGNet 的前 10 层用于为输入图像创建特征图。

（2）阶段 1：使用 2 分支多阶段 CNN。其中，第一个分支预测一组身体部位位置（如肘部、膝盖等）的二维置信度图。置信度图是灰度图像，在某个身体部位的可能性很高的位置具有很高的值。对于 18 点模型，输出的前 19 个矩阵对应于置信度图。第二个分支预测一组二维向量

场（L）的部分亲缘关系（PAF），它编码了各部分（关键点）之间的关联程度。第 20~57 个矩阵是 PAF 矩阵。

从这里（http://posefs1.perception.cs.cmu.edu/OpenPose/models/pose/coco/pose_iter_440000.caffemodel）下载模型。下载权重文件后，将其放在"pose/coco/"文件夹中。

1）从图像生成输出

加载网络的代码如下：

```
protoFile = "pose/coco/pose_deploy_linevec.prototxt"
weightsFile = "pose/coco/pose_iter_440000.caffemodel"
net = cv2.dnn.readNetFromCaffe(protoFile, weightsFile)
```

加载图像并创建输入 blob：

```
image1 = cv2.imread("group.jpg")
# Fix the input Height and get the width according to the Aspect Ratio
inHeight = 368
inWidth = int((inHeight/frameHeight)*frameWidth)

inpBlob = cv2.dnn.blobFromImage(image1, 1.0 / 255, (inWidth, inHeight),
                  (0, 0, 0), swapRB=False, crop=False)
```

正向通过网络（执行预测），代码如下：

```
net.setInput(inpBlob)
output = net.forward()
```

首先将输出调整为与输入相同的大小，然后检查鼻子关键点对应的置信度图。还可以使用 cv2.addWeighted()函数对图像上的 probMap 进行 alpha 混合。

```
i = 0
probMap = output[0, i, :, :]
probMap = cv2.resize(probMap, (frameWidth, frameHeight))

plt.imshow(cv2.cvtColor(image1, cv2.COLOR_BGR2RGB))
plt.imshow(probMap, alpha=0.6)
```

2）检测关键点

对于一个人来说，只需找到置信度图的最大值，就很容易找到每个关键点的位置，但是对于多人场景不能这样做。

对于每个关键点，我们将阈值（在这里是 0.1）应用于置信度图。Python 代码如下：

```
mapSmooth = cv2.GaussianBlur(probMap,(3,3),0,0)
mapMask = np.uint8(mapSmooth>threshold)
```

这里给出了一个矩阵，其中包含与关键点对应的区域中的斑点。为了找到关键点的确切位置，需要找到每个斑点的最大值。执行以下操作：

（1）找到与关键点对应区域的所有轮廓。

（2）为这个区域创建一个蒙版。

（3）通过将 probMap 与此蒙版相乘来提取该区域的 probMap。

（4）找到该区域的局部最大值。这是为每个轮廓（关键点区域）完成的。

Python 代码如下：

```
#find the blobs
```

```
_, contours, _ = cv2.findContours(mapMask, cv2.RETR_TREE, cv2.CHAIN_APPROX_SIMPLE)

#for each blob find the maxima
for cnt in contours:
    blobMask = np.zeros(mapMask.shape)
    blobMask = cv2.fillConvexPoly(blobMask, cnt, 1)
    maskedProbMap = mapSmooth * blobMask
    _, maxVal, _, maxLoc = cv2.minMaxLoc(maskedProbMap)
    keypoints.append(maxLoc + (probMap[maxLoc[1], maxLoc[0]],))
```

保存每个关键点的 x、y 坐标和概率分数。还为找到的每个关键点分配一个 ID。

3）查找有效对

有效对是连接两个关键点的身体部位，属于同一个人。找到有效对的一种简单方法是找到一个关节与所有可能的其他关节之间的最小距离。

这种方法可能不适用于所有配对，特别是当图像包含太多人或有部分遮挡时。

对于每个身体部位对，执行以下操作：

（1）取属于一对的关键点，并将它们放在单独的列表中（candA 和 candB）。candA 中的每个点都将连接到 candB 中的某个点。

Python 代码如下：

```
pafA = output[0, mapIdx[k][0], :, :]
pafB = output[0, mapIdx[k][1], :, :]
pafA = cv2.resize(pafA, (frameWidth, frameHeight))
pafB = cv2.resize(pafB, (frameWidth, frameHeight))

# Find the keypoints for the first and second limb
candA = detected_keypoints[POSE_PAIRS[k][0]]
candB = detected_keypoints[POSE_PAIRS[k][1]]
```

（2）找到连接所考虑的两点的单位向量。这给出了连接它们的线的方向。

Python 代码如下：

```
d_ij = np.subtract(candB[j][:2], candA[i][:2])
norm = np.linalg.norm(d_ij)
if norm:
    d_ij = d_ij / norm
```

（3）在连接两个点的线上创建一个包含 10 个插值点的数组。

Python 代码如下：

```
# Find p(u)
interp_coord = list(zip(np.linspace(candA[i][0], candB[j][0], num=n_interp_samples),
                    np.linspace(candA[i][1], candB[j][1], num=n_interp_samples)))
# Find L(p(u))
paf_interp = []
for k in range(len(interp_coord)):
    paf_interp.append([pafA[int(round(interp_coord[k][1])), int(round(interp_coord[k][0]))],
                pafB[int(round(interp_coord[k][1])), int(round(interp_coord[k][0]))] ])
```

（4）取这些点上的 PAF 与单位向量 d_ij 之间的点积。

Python 代码如下：

```
# Find E
paf_scores = np.dot(paf_interp, d_ij)
avg_paf_score = sum(paf_scores)/len(paf_scores)
```

(5) 如果 70%的点满足标准,则称该配对有效。

Python 代码如下:

```
# Check if the connection is valid
# If the fraction of interpolated vectors aligned with PAF is higher then threshold
# -> Valid Pair
if ( len(np.where(paf_scores > paf_score_th)[0]) / n_interp_samples ) > conf_th :
    if avg_paf_score > maxScore:
        max_j = j
        maxScore = avg_paf_score
```

4) 组装个人关键点

现在已经将所有关键点组合成对,可以将共享相同部位检测候选的对组合成多人的全身姿势。下面来看看它是如何通过代码完成的。

(1) 首先创建空列表来存储每个人的关键点,然后检查每一对——检查各对的 partA 是否已经存在于任何列表中。如果它存在,则意味着关键点属于这个列表,并且各对的 partB 也应该属于这个人。因此,将此对的 partB 添加到找到 partA 的列表中。代码如下:

```
for j in range(len(personwiseKeypoints)):
    if personwiseKeypoints[j][indexA] == partAs[i]:
        person_idx = j
        found = 1
        break

if found:
    personwiseKeypoints[person_idx][indexB] = partBs[i]
```

(2) 如果 partA 不存在于任何列表中,则意味着该对属于不在列表中的新人,因此创建一个新列表。代码如下:

```
elif not found and k < 17:
    row = -1 * np.ones(19)
    row[indexA] = partAs[i]
    row[indexB] = partBs[i]
```

检查每个人并在输入图像上绘制骨架。代码如下:

```
for i in range(17):
    for n in range(len(personwiseKeypoints)):
        index = personwiseKeypoints[n][np.array(POSE_PAIRS[i])]
        if -1 in index:
            continue
        B = np.int32(keypoints_list[index.astype(int), 0])
        A = np.int32(keypoints_list[index.astype(int), 1])
        cv2.line(frameClone, (B[0], A[0]), (B[1], A[1]), colors[i], 2, cv2.LINE_AA)

cv2.imshow("Detected Pose" , frameClone)
cv2.waitKey(0)
```